Separation of Variables for Partial Differential Equations

An Eigenfunction Approach

Studies in Advanced Mathematics

Separation of Variables for Partial Differential Equations

An Eigenfunction Approach

George Cain

Georgia Institute of Technology
Atlanta, Georgia, USA

Gunter H. Meyer

Georgia Institute of Technology
Atlanta, Georgia, USA

CRC Press
Taylor & Francis Group
Boca Raton London New York

CRC Press is an imprint of the
Taylor & Francis Group, an **informa** business

A CHAPMAN & HALL BOOK

Chapman & Hall/CRC Press
Taylor & Francis Group
6000 Broken Sound Parkway NW, Suite 300
Boca Raton, FL 33487-2742

© 2006 by Taylor & Francis Group, LLC
CRC Press is an imprint of Taylor & Francis Group, an Informa business

First issued in paperback 2019

No claim to original U.S. Government works

ISBN-13: 978-0-367-44643-7 (pbk)
ISBN-13: 978-1-58488-420-0 (hbk)

Visit the Taylor & Francis Web site at
http://www.taylorandfrancis.com

and the CRC Press Web site at
http://www.crcpress.com

Library of Congress Card Number 2005051950

Library of Congress Cataloging-in-Publication Data

Cain, George L.
  Separation of variables for partial differential equations : an eigenfunction approach / George Cain, Gunter H. Meyer.
        p. cm. -- (Studies in advanced mathematics)
  Includes bibliographical references and index.
  ISBN 1-58488-420-7 (alk. paper)
  1. Separation of variables. 2. Eigenfunctions. I. Meyer, Gunter H. II. Title. III. Series.

QA377.C247 2005
515'.353--dc22   2005051950

Acknowledgments

We would like to thank our editor Sunil Nair for welcoming the project and for his willingness to stay with it as it changed its scope and missed promised deadlines.

We also wish to express our gratitude to Ms. Annette Rohrs of the School of Mathematics of Georgia Tech who transformed decidedly low-tech scribbles into a polished manuscript. Without her talents, and patience, we would not have completed the book.

Preface

Separation of variables is a solution method for partial differential equations. While its beginnings date back to work of Daniel Bernoulli (1753), Lagrange (1759), and d'Alembert (1763) on wave motion (see [2]), it is commonly associated with the name of Fourier (1822), who developed it for his research on conductive heat transfer. Since Fourier's time it has been an integral part of engineering mathematics, and in spite of its limited applicability and heavy competition from numerical methods for partial differential equations, it remains a well-known and widely used technique in applied mathematics.

Separation of variables is commonly considered an analytic solution method that yields the solution of certain partial differential equations in terms of an infinite series such as a Fourier series. While it may be straightforward to write formally the series solution, the question in what sense it solves the problem is not readily answered without recourse to abstract mathematical analysis. A modern treatment focusing in part on the theoretical underpinnings of the method and employing the language and concepts of Hilbert spaces to analyze the infinite series may be found in the text of MacCluer [15]. For many problems the formal series can be shown to represent an analytic solution of the differential equation. As a tool of analysis, however, separation of variables with its infinite series solutions is not needed. Other mathematical methods exist which guarantee the existence and uniqueness of a solution of the problem under much more general conditions than those required for the applicability of the method of separation of variables.

In this text we mostly ignore infinite series solutions and their theoretical and practical complexities. We concentrate instead on the first N terms of the series which are all that ever are computed in an engineering application. Such a partial sum of the infinite series is an approximation to the analytic solution of the original problem. Alternatively, it can be viewed as the exact analytic solution of a new problem that approximates the given problem. This is the point of view taken in this book.

Specifically, we view the method of separation of variables in the following context: mathematical analysis applied to the given problem guarantees the existence and uniqueness of a solution u in some infinite dimensional vector space of functions X, but in general provides no means to compute it. By modifying the problem appropriately, however, an approximating problem results which has a computable closed form solution u_N in a subspace M of X. If M is suitably chosen, then u_N is a good approximation to the unknown solution u. As we shall see, M will be defined such that u_N is just the partial sum of the first N terms of the infinite series traditionally associated with the method of separation of variables.

The reader may recognize this view as identical to the setting of the finite element, collocation, and spectral methods that have been developed for the numerical solution of differential equations. All these methods differ in how the subspace M is chosen and in what sense the original problem is approximated. These choices dictate how hard it is to compute the approximate solution u_N and how well it approximates the analytic solution u.

Given the almost universal applicability of numerical methods for the solution of partial differential equations, the question arises whether separation of variables with its severe restrictions on the type of equation and the geometry of the problem is still a viable tool and deserves further exposition. The existence of this text reflects our view that the method of separation of variables still belongs to the core of applied mathematics. There are a number of reasons.

Closed form (approximate) solutions show structure and exhibit explicitly the influence of the problem parameters on the solution. We think, for example, of the decomposition of wave motion into standing waves, of the relationship between driving frequency and resonance in sound waves, of the influence of diffusivity on the rate of decay of temperature in a heated bar, or of the generation of equipotential and stream lines for potential flow. Such structure and insight are not readily obtained from purely numerical solutions of the underlying differential equation. Moreover, optimization, control, and inverse problems tend to be easier to solve when an analytic representation of the (approximate) solution is available. In addition, the method is not as limited in its applicability as one might infer from more elementary texts on separation of variables. Approximate solutions are readily computable for problems with time-dependent data, for diffusion with convection and wave motion with dissipation, problems seldom seen in introductory textbooks. Even domain restrictions can sometimes be overcome with embedding and domain decomposition techniques. Finally, there is the class of singularly perturbed and of higher dimensional problems where numerical methods are not easily applied while separation of variables still yields an analytic approximate solution.

Our rationale for offering a new exposition of separation of variables is then twofold. First, although quite common in more advanced treatments (such as [15]), interpreting the separation of variables solution as an eigenfunction expansion is a point of view rarely taken when introducing the method to students. Usually the formalism is based on a product solution for the partial differential equation, and this limits the applicability of the method to homogeneous partial differential equations. When source terms do appear, then a reformulation of problems for the heat and wave equation with the help of Duhamel's superposition principle and an approximation of the source term in the potential equation with the help of an eigenfunction approximation become necessary. In an exposition based from the beginning on an eigenfunction expansion, the presence of source terms in the differential equation is only a technical, but not a conceptual complication, regardless of the type of equation under consideration. A concise algorithmic approach results.

Equally important to us is the second reason for a new exposition of the method of separation of variables. We wish to emphasize the power of the method by solving a great variety of problems which often go well beyond the usual textbook examples. Many of the applications ask questions which are not as easily resolved with numerical methods as with analytic approximate solutions. Of course, evaluation of these approximate solutions usually relies on numerical methods to integrate, solve linear systems or nonlinear equations, and to find values of special functions, but these methods by now may be considered universally available "black boxes." We are, however, mindful of the gap between the concept of a solution in principle and a demonstrably computable solution and try to convey our experience with how well the eigenfunction approach actually solves the sample problems.

The method of separation of variables from a spectral expansion view is presented in nine chapters.

Chapter 1 collects some background information on the three dominant equations of this text, the potential equation, the heat equation, and the wave equation. We refer to these results when applying and analyzing the method of separation of variables.

Chapter 2 contains a discussion of orthogonal projections which are used time and again to approximate given data functions in a specified finite-dimensional but parameter-dependent subspace.

Chapter 3 introduces the subspace whose basis consists of the eigenfunctions of a so-called Sturm–Liouville problem associated with the application under consideration. These are the eigenfunctions of the title of this text. We cite results from the Sturm–Liouville theory and provide a table of eigenvalues and eigenfunctions that arise in the method of separation of variables.

Chapter 4 treats the case in which the eigenfunctions are sine and cosine functions with a common period. In this case the projection into the subspace is closely related to the Fourier series representation of the data functions. Precise information about the convergence of the Fourier series is known. We cite those results which are helpful later on for the application of separation of variables.

Chapter 5 constitutes the heart of the text. We consider a partial differential equation in two independent variables with a source term and subject to boundary and initial conditions. We give the algorithm for approximating such a problem and for solving it in a finite-dimensional space spanned by eigenfunctions determined by the "spacial part" of the equation and its boundary conditions. We illustrate in broad outline the application of this approach to the heat, wave, and potential equations.

Chapter 6 gives an expansive exposition of the algorithm for the one-dimensional heat equation. It contains many worked examples with comments on the numerical performance of the method, and concludes with a rudimentary analysis of the error in the approximate solution.

Chapter 7 parallels the previous chapter but treats the wave equation.

Chapter 8 deals with the potential equation. It describes how one can precondition the data of problems with smooth solutions in order not to introduce artificial discontinuities into the separation of variables solution. We solve potential problems with various boundary conditions and conclude with a calculation of eigenfunctions for the two-dimensional Laplacian.

Chapter 9 uses the eigenfunctions of the preceding chapter to find eigenfunction expansion solutions of two- and three-dimensional heat, wave, and potential equations.

This text is written for advanced undergraduate and graduate students in science and engineering with previous exposure to a course in engineering mathematics, but not necessarily separation of variables. Basic prerequisites beyond calculus are familiarity with linear algebra, the concept of vector spaces of functions, norms and inner products, the ability to solve linear inhomogeneous first and second order ordinary differential equations, and some contact with practical applications of partial differential equations.

The book contains more material than can (and should) be taught in a course on separation of variables. We have introduced the eigenfunction approach to our own students based on an early version of this text. We covered parts of Chapters 2–4 to lay the groundwork for an extensive discussion of Chapter 5. The remainder of the term was filled by working through selected examples involving the heat, wave, and potential equation. We believe that by term's end the students had an appreciation that they could solve realistic problems. Since we view Chapters 2–5 as suitable for teaching separation of variables, we have included exercises to help deepen the reader's understanding of the eigenfunction approach. The examples of Chapters 6–8 and their exercise sets generally lend themselves for project assignments.

This text will put a bigger burden on the instructor to choose topics and guide students than more elementary texts on separation of variables that start with product solutions. The instructor who subscribes to the view put forth in Chapter 5 should find this text workable. The more advanced applications, such as interface, inverse, and multidimensional problems, as well as the the more theoretical topics require more mathematical sophistication and may be skipped without breaking continuity.

The book is also meant to serve as a reference text for the method of separation of variables. We hope the many examples will guide the reader in deciding whether and how to apply the method to any given problem. The examples should help in interpreting computed solutions, and should give insight into those cases in which formal answers are useless because of lack of convergence or unacceptable oscillations. Chapters 1 and 9 are included to support the reference function. They do not include exercises.

We hasten to add that this text is not a complete reference book. We do not attempt to characterize the equations and coordinate systems where a separation of variables is applicable. We do not even mention the various coordinate systems (beyond cartesian, polar, cylindrical, and spherical) in which the Laplacian is separable. We have not scoured the literature for new and innovative applications of separation of variables. Moreover, the examples we do include are often meant to show structure rather than represent reality because in general little attention is given to the proper scaling of the equations.

There does not appear to exist any other source that could serve as a practical reference book for the practicing engineer or scientist. We hope this book will alert the reader that separation of variables has more to offer than may be apparent from elementary texts.

Finally, this text does not mention the implementation of our formulas and calculations on the computer, or do we provide numerical algorithms or programs. Yet the text, and in particular our numerical examples, could not have been presented without access to symbolic and numerical packages such as Maple, Mathematica, and Matlab. We consider our calculations and the graphical representation of their results routine and well within the competence of today's students and practitioners of science and engineering.

Contents

Chapter 1

Potential, Heat, and Wave Equation

This chapter provides a quick look into the vast field of partial differential equations. The main goal is to extract some qualitative results on the three dominant equations of mathematical physics, the potential, heat, and wave equation on which our attention will be focused throughout this text.

1.1 Overview

When processes that change smoothly with two or more independent variables are modeled mathematically, then partial differential equations arise. Most common are second order equations of the general form

$$\mathcal{L}u \equiv \sum_{i,j=1}^{M} a_{ij} u_{x_i x_j} + \sum_{i=1}^{M} b_i u_{x_i} + cu = F, \qquad \vec{x} \in D \subset \mathbb{R}_M \qquad (1.1)$$

where the coefficients and the source term may depend on the independent variables $\{x_1, \ldots, x_M\}$, on u, and on its derivatives. D is a given set in $\mathbb{R}_M$ (whose boundary will be denoted by ∂D). The equation may reflect conservation and balance laws, empirical relationships, or may be purely phenomenological. Its solution is used to explain, predict, and control processes in a bewildering array of applications ranging from heat, mass, and fluid flow, migration of biological species, electrostatics, and molecular vibration to mortgage banking.

In (1.1) $\mathcal{L}$ is known as a partial differential operator that maps a smooth function u to the function $\mathcal{L}u$. Throughout this text a smooth function denotes a function with as many continuous derivatives as are necessary to carry out the operations to which it is subjected. $\mathcal{L}u = F$ is the equation to be solved.

Given a partial differential equation and side constraints on its solution, typically initial and boundary conditions, it becomes a question of mathematical analysis to establish whether the problem has a solution, whether the solution is unique, and whether the solution changes continuously with the data of the problem. If that is the case, then the given problem for (1.1) is said to be well posed; if not then it is ill posed. We note here that the data of the problem are the coefficients of $\mathcal{L}$, the source term F, any side conditions imposed on u, and the shape of D. However, dependence on the coefficients and on the shape of D will be ignored. Only continuous dependence with respect to the source term and the side conditions will define well posedness for our purposes.

The technical aspects of in what sense a function u solves the problem and in what sense it changes with the data of the problem tend to be abstract and complex and constitute the mathematical theory of partial differential equations (e.g., [5]). Such theoretical studies are essential to establish that equation (1.1) and its side conditions are a consistent description of the processes under consideration and to characterize the behavior of its solution. Outside mathematics the validity of a mathematical model is often taken on faith and its solution is assumed to exist on "physical grounds." There the emphasis is entirely on solving the equation, analytically if possible, or approximately and numerically otherwise. Approximate solutions are the subject of this text.

1.2   Classification of second order equations

The tools for the analysis and solution of (1.1) depend on the structure of the coefficient matrix

$$\mathcal{A} = \{a_{ij}\}$$

in (1.1). By assuming that $u_{x_i x_j} = u_{x_j x_i}$ we can always write $\mathcal{A}$ in such a way that it is symmetric. For example, if the equation which arises in modeling a process is

$$\mathcal{L}u \equiv u_{x_1 x_1} + 2u_{x_1 x_2} + u_{x_2 x_2} = 0,$$

then it will be rewritten as

$$\mathcal{L}u \equiv u_{x_1 x_1} + u_{x_1 x_2} + u_{x_2 x_1} + u_{x_2 x_2} = 0$$

so that

$$\mathcal{A} = \begin{pmatrix} 1 & 1 \\ 1 & 1 \end{pmatrix}.$$

We can now introduce three broad classes of differential equations.

Definition The operator $\mathcal{L}$ given by

$$\mathcal{L}u \equiv \sum_{i,j=1}^{M} a_{ij} u_{x_i x_j} + \sum_{i=1}^{M} b_i u_{x_i} + cu$$

is

i) Elliptic at $\vec{x} = (x_1, \ldots, x_M)$ if all eigenvalues of the symmetric matrix $\mathcal{A}$ are nonzero and have the same algebraic sign,

ii) Hyperbolic at $\vec{x}$ if all eigenvalues of $\mathcal{A}$ are nonzero and one has a different algebraic sign from all others,

iii) Parabolic at $\vec{x}$ if $\mathcal{A}$ has a zero eigenvalue.

If $\mathcal{A}$ depends on u and its derivatives, then $\mathcal{L}$ is elliptic, etc. at a given point relative to a specific function u. If the operator $\mathcal{L}$ is elliptic at a point then (1.1) is an elliptic equation at that point. (As mnemonic we note that for $M = 2$ the level sets of

$$\left\langle \mathcal{A}\begin{pmatrix} x_1 \\ x_2 \end{pmatrix}, \begin{pmatrix} x_1 \\ x_2 \end{pmatrix} \right\rangle = \text{constant},$$

where $\langle \vec{x}, \vec{y} \rangle$ denotes the dot product of $\vec{x}$ and $\vec{y}$, are elliptic, hyperbolic, and parabolic under the above conditions on the eigenvalues of $\mathcal{A}$). The lower order terms in (1.1) do not affect the type of the equation, but in particular applications they can dominate the behavior of the solution of (1.1).

Each class of equations has its own admissible side conditions to make (1.1) well posed, and all solutions of the same class have, broadly speaking, common characteristics. We shall list some of them for the three dominant equations of mathematical physics: Laplace's equation, the heat equation, and the wave equation.

1.3 Laplace's and Poisson's equation

The most extensively studied example of an elliptic equation is Laplace's equation

$$\mathcal{L}u \equiv \nabla \cdot \nabla u = 0$$

which arises in potential problems, steady-state heat conduction, irrotational flow, minimal surface problems, and myriad other applications. The operator $\mathcal{L}u$ is known as the Laplacian and is generally denoted by

$$\mathcal{L}u \equiv \nabla \cdot \nabla u \equiv \nabla^2 u \equiv \Delta u.$$

The last form is common in the mathematical literature and will be used consistently throughout this text. The Laplacian in cartesian coordinates

$$\Delta u \equiv \sum_{i=1}^{M} u_{x_i x_i}$$

assumes the forms

i) In polar coordinates (r, θ)

$$\Delta u \equiv u_{rr} + \frac{1}{r} u_r + \frac{1}{r^2} u_{\theta\theta},$$

ii) In cylindrical coordinates (r, θ, z)

$$\Delta u \equiv u_{rr} + \frac{1}{r} u_r + \frac{1}{r^2} u_{\theta\theta} + u_{zz},$$

iii) In spherical coordinates (r, θ, ϕ)

$$\Delta u \equiv u_{rr} + \frac{2}{r} u_r + \frac{1}{r^2 \sin \phi} (\sin \phi u_\phi)_\phi + \frac{1}{r^2 \sin^2 \phi} u_{\theta\theta}.$$

For special applications other coordinate systems may be more advantageous and we refer to the literature (see, e.g., [13]) for the representation of Δu in additional coordinate systems. For cartesian coordinates we shall use the common notation

$$\vec{x} = (x, y), \qquad \vec{x} = (x, y, z)$$

for $\vec{x} \in \mathbb{R}_2$ and $\mathbb{R}_3$, respectively.

The generalization of Laplace's equation to

$$\Delta u = F \tag{1.2}$$

for a given source term F is known as Poisson's equation. It will be the dominant elliptic equation in this text.

In general equation (1.2) is to be solved for $\vec{x} \in D$ where D is an open set in $\mathbb{R}_M$. For the applications in this text D will usually be a bounded set with a sufficiently smooth boundary ∂D. On ∂D the solution u has to satisfy boundary conditions. We distinguish between three classes of boundary data for (1.2) and its generalizations.

i) The Dirichlet problem (also known as a problem of the first kind)

$$u = g(\vec{x}), \qquad \vec{x} \in \partial D.$$

ii) The Neumann problem (also known as a problem of the second kind)

$$\frac{\partial u}{\partial n} = g(\vec{x}), \qquad \vec{x} \in \partial D$$

where $\frac{\partial u}{\partial n} = \nabla u \cdot \vec{n}(\vec{x})$ is the normal derivative of u, i.e., the directional derivative of u in the direction of the outward unit normal $\vec{n}(\vec{x})$ to D at $\vec{x} \in \partial D$.

iii) The Robin problem (also known as a problem of the third kind):

$$\alpha_1 \frac{\partial u}{\partial n} + \alpha_2 u = g(\vec{x}), \qquad \vec{x} \in \partial D,$$

where, at least in this text, α_1 and α_2 are piecewise nonnegative constants.

We shall call general boundary value problems for (1.2) potential problems. The differential equation and the boundary conditions are called homogeneous if the function $u(\vec{x}) \equiv 0$ can satisfy them. For example, (1.2) is homogeneous if $F(\vec{x}) \equiv 0$, and the boundary data are homogeneous if $g(\vec{x}) \equiv 0$.

If D is a bounded open set which has a well-defined outward normal at every point of ∂D, then for continuous functions F on D and g on ∂D a classical solution of these three problems is a function u which is twice continuously differentiable in D and satisfies (1.2) at every point of D. In addition, the classical solution of the Dirichlet problem is required to be continuous on the closed set $\bar{D} = D \cup \partial D$ and equal to g on ∂D. For problems of the second and third kind the classical solution also needs continuous first derivatives on $\bar{D}$ in order to satisfy the given boundary condition on ∂D.

The existence of classical solutions is studied in great generality in [6]. It is known that for continuous F and g and smooth ∂D the Dirichlet problem has a classical solution, and that the Robin problem has a classical solution whenever F and g are continuous and

$$\alpha_1 \alpha_2 > 0.$$

A classical solution for the Neumann problem is known to exist for continuous F and g provided

$$\int_D F(\vec{x})d\vec{x} = \int_{\partial D} g(\vec{s})d\vec{s}.$$

Why this compatibility condition arises is discussed below.

Considerable effort has been devoted in the mathematical literature to extending these existence results to domains with corners and edges where the normal is not defined, and to deriving analogous results when F and g (and the coefficients in (1.1)) are not necessarily continuous. Classical solutions no longer exist but so-called weak solutions can be defined which solve integral equations derived from (1.1) and the boundary conditions. This general existence theory is also presented in [6].

In connection with separation of variables we shall be concerned only with bounded elementary domains like rectangles, wedges, cylinders, balls, and shells where the boundaries are smooth except at isolated corners and edges. At such points the normal is not defined. Likewise, isolated discontinuities in the data functions F and g may occur. We shall assume throughout the book (with optimism, or on physical grounds) that the given problems have weak solutions which are smooth and satisfy the differential equation and boundary conditions at all points where the data are continuous.

Our separation of variables solution will be an approximation to the analytic solution found by smoothing the data F and g. Such an approximation can only be meaningful if the solution of the original boundary value problem depends continuously on the data, in other words, if the boundary value problem is well posed. We shall examine this question for the Dirichlet and Neumann problem.

Dirichlet problem: The Dirichlet problem for Poisson's equation is the most thoroughly studied elliptic boundary value problem. We shall assume that F is continuous for $\vec{x} \in D$ and g is continuous for $\vec{x} \in \partial D$ so that the problem

$$\Delta u = F(\vec{x}), \qquad \vec{x} \in D$$

$$u = g(\vec{x}), \qquad \vec{x} \in \partial D$$

has a classical solution. Any approximating problem formulated to solve the Dirichlet problem analytically should likewise have a classical solution. As discussed in Chapter 8, this may require preconditioning the problem before applying separation of variables.

We remark that especially for establishing the existence of a solution it often is advantageous to split the solution

$$u = u_1 + u_2$$

where

$$\Delta u_1 = 0, \qquad \vec{x} \in D$$
$$u_1 = g(\vec{x}), \qquad \vec{x} \in \partial D$$
$$\Delta u_2 = F(\vec{x}), \qquad \vec{x} \in D$$
$$u_2 = 0, \qquad \vec{x} \in \partial D$$

because different mathematical tools are available for Laplace's equation with nonzero boundary data and for Poisson's equation with zero boundary data (which, in an abstract sense, has a good deal in common with the matrix problem $Au = b$). Splittings of this type will be used routinely in Chapter 8. Of course, if g is defined and continuous on all of $\bar{D}$ and twice continuously differentiable in the open set D, then it is usually advantageous to introduce the new function

$$w \equiv u - g$$

and solve the Dirichlet problem

$$\Delta w = F - \Delta g, \qquad \vec{x} \in D$$

$$w = 0, \qquad \vec{x} \in \partial D$$

without splitting.

Given a classical solution we now wish to show that it depends continuously on F and g. To give meaning to this phrase we need to be able to measure change in the functions F and g. Here this will be done with respect to the so-called supremum norm. We recall from analysis that for any function G defined on a set $D \subset R_M$

$$\sup_{\vec{y} \in D} |G(\vec{y})| = \text{least upper bound on the set of values } \{|G(\vec{y})| : \vec{y} \in D\}.$$

A common notation is

$$\|G\| = \sup_{\vec{y} \in D} |G(\vec{y})|$$

which is called the supremum norm of G and which is just one example of the concept of a norm discussed in Chapter 2.

If $\bar{D}$ is a closed and bounded set and G is continuous in $\bar{D}$, then G must take on its maximum and minimum on $\bar{D}$ so that

$$\|G\| = \sup_{\vec{y} \in \bar{D}} |G(\vec{y})| = \max_{\vec{y} \in \bar{D}} |G(\vec{y})|.$$

Continuous dependence of a classical solution (with respect to the supremum norm) is given if for every $\epsilon > 0$, there is a $\delta > 0$ such that

$$\|u\| \leq \epsilon$$

whenever

$$\|F\| + \|g\| \leq \delta.$$

Here

$$\|u\| = \max_{\vec{x} \in \bar{D}} |u(\vec{x})|, \quad \|F\| = \sup_{\vec{x} \in D} |F(\vec{x})|, \quad \text{and} \quad \|g\| = \max_{\vec{x} \in \partial D} |g(\vec{x})|.$$

Continuous dependence on the data in this sense, and uniqueness follow from the maximum principle for elliptic equations. Since it is used later on and always provides a quick check on computed or approximate solutions of the Dirichlet problem, and since it is basically just the second derivative test of elementary calculus, we shall briefly discuss it here.

Theorem 1.1 *The maximum principle*
Let u be a smooth solution of Poisson's equation

$$\Delta u = F, \qquad \vec{x} \in D. \tag{1.3}$$

Assume that

$$F > 0 \quad \text{for all } \vec{x} \in D.$$

Then $u(\vec{x})$ cannot assume a relative maximum in D.

Proof. If u has a relative maximum at some $\vec{x}_0 \in D$, then the second derivative test requires that $u_{x_i x_i}(\vec{x}_0) \leq 0$ for all i which would contradict $\Delta u(\vec{x}_0) = F(\vec{x}_0) > 0$.

We note that if $\bar{D}$ is bounded, then a classical solution u of the Dirichlet problem must assume a maximum at some point in $\bar{D}$. Since this point cannot lie in D, it must lie on ∂D. Hence $u(\vec{x}) \leq \max_{\vec{y} \in \partial D} g(\vec{y})$ for all $\vec{x} \in \bar{D}$. We also note that if $F < 0$, then $-u$ satisfies the above maximum principle which translates into $u(\vec{x}) \geq \min_{\vec{y} \in \partial D} g(\vec{y})$.

Stronger statements, extensions to the general elliptic equation (1.1), and more general boundary conditions may be found in most texts on partial differential equations (see, e.g., [6]).

Theorem 1.2 *A solution of the Dirichlet problem depends continuously on the data F and g.*

Proof. Suppose that D is such that $a < x_k < b$ for some k, $1 < k < M$ where a and b denote finite lower and upper bounds on the kth coordinate of $\vec{x} \in D$. Define the function

$$\phi(\vec{x}) = (\|F\| + \epsilon) \frac{(x_k - a)^2}{2}$$

where $\epsilon > 0$ is arbitrary and where $\|F\| = \sup_{\vec{x} \in D} |F(\vec{x})|$ and $\|g\| = \max_{\vec{x} \in \partial D} |g(\vec{x})|$ are assumed to be finite. Then

$$\Delta \left[\phi(\vec{x}) \pm u(\vec{x})\right] \geq \epsilon > 0$$

and by the maximum principle $[\phi \pm u]$ is bounded above by its value on ∂D so that

$$\pm u(\vec{x}) \leq \phi(\vec{x}) \pm u(\vec{x}) \leq (\|F\| + \epsilon)\frac{(b-a)^2}{2} + \|g\|.$$

Since ϵ is arbitrary, it follows that for all $\vec{x} \in D$

$$|u(\vec{x})| \leq \|F\|\frac{(b-a)^2}{2} + \|g\|$$

which implies

$$\|u\| \leq \|F\|\frac{(b-a)^2}{2} + \|g\|. \tag{1.4}$$

This inequality establishes continuous dependence since $\|F\| \to 0$ and $\|g\| \to 0$ imply that $\max_{\bar{D}} |u(\vec{x})| \to 0$.

Corollary 1.3 *The solution of the Dirichlet problem is unique.*

Proof. The difference between two solutions satisfies the Dirichlet problem

$$\Delta u = 0 \quad \text{in } D$$
$$u = 0 \quad \text{on } \partial D$$

which by Theorem 1.2 has only the zero solution.

Corollary 1.4 *Let $F \geq 0$. Then the solution of the Dirichlet problem assumes its maximum on ∂D.*

Proof. Let $\epsilon > 0$ be arbitrary. Then the solution u_ϵ of

$$\Delta u = F + \epsilon \quad \text{in } D$$
$$u = g \quad \text{on } \partial D$$

assumes its maximum on ∂D by Theorem 1.1. By Theorem 1.2

$$|u_\epsilon(x) - u(x)| \leq \epsilon \frac{(b-a)^2}{2}.$$

Since ϵ is arbitrary, u cannot exceed $\max_{\vec{y} \in \partial D} g(\vec{y})$ by a nonzero amount at any point in $\bar{D}$; hence $u(\vec{x}) \leq \max_{\vec{y} \in \partial D} g(\vec{y})$.

Similarly, if $F \leq 0$, then u assumes its minimum on ∂D. Consequently, the solution of Laplace's equation

$$\Delta u = 0, \qquad \vec{x} \in D$$

must assume its maximum and minimum on ∂D.

Looking ahead, we see that in Chapter 8 the solution u of the Dirichlet problem for (1.2) will be approximated by the computable solution u_N of a related Dirichlet problem

$$\Delta u = F_N, \qquad \vec{x} \in D \qquad (1.5)$$
$$u = g_N, \qquad \vec{x} \in \partial D.$$

The existence of u_N is given because it will be found explicitly. Uniqueness of the solution guarantees that no other solution of (1.5) exists. It only remains to establish in what sense u_N approximates the analytic solution u. But it is clear from Theorem 1.2 that for all $x \in \bar{D}$

$$|u_N(\vec{x}) - u(\vec{x})| \le K \|F_N - F\| + \|g_N - g\| \qquad (1.6)$$

where the constant K depends only on the geometry of D. Thus the error in the approximation depends on how well F_N and g_N approximate the given data F and g. These issues are discussed in Chapters 3 and 4.

When the Dirichlet problem does not have a classical solution because the data are not smooth, then continuous dependence for weak solutions must be established. It generally is possible to show that if the data tend to zero in a mean square sense, then the weak solution of the Dirichlet problem tends to zero in a mean square sense. This translates into mean square convergence of u_N to u. The analysis of such problems becomes demanding and we refer to [6] for details. A related result for the heat equation is discussed in Section 6.2.

Neumann problem: In contrast to the Dirichlet problem, the Neumann problem for Poisson's equation is not well posed because if u is a solution then $u + c$ for any constant c is also a solution, hence a solution is not unique. But there may not be a solution at all if the data are inconsistent. Suppose that u is a solution of the Neumann problem; then it follows from the divergence theorem that

$$\int_D F(\vec{x})d\vec{x} = \int_D \Delta u\, d\vec{x} = \int_D \nabla \cdot \nabla u\, d\vec{x}$$
$$= \int_{\partial D} \nabla u(\vec{s}) \cdot \vec{n}(\vec{s})d\vec{s} = \int_{\partial D} g(\vec{s})d\vec{s},$$

where $\vec{n}$ is the outward unit normal on ∂D. Hence a necessary condition for the existence of a solution is the compatibility condition

$$\int_D F(\vec{x})d\vec{x} = \int_{\partial D} g(\vec{s})d\vec{s}. \qquad (1.7)$$

If we interpret the Neumann problem for Poisson's equation as a (scaled) steady-state heat transfer problem, then this compatibility condition simply states that the energy generated (or destroyed) in D per unit time must be balanced exactly by the energy flux across the boundary of D. Were this not the case D would warm up or cool down and could not have a steady-state temperature. Equation

(1.7) implies that the solution does not change continuously with the data since F and g cannot be changed independently. If, however, (1.7) does hold, then the Neumann problem is known to have a classical solution which is unique up to an additive constant. In practice the solution is normalized by assigning a value to the additive constant, e.g., by requiring $u(\vec{x}_0) = 1$ for some fixed $x_0 \in \bar{D}$. The Neumann problems arise frequently in applications and can be solved with separation of variables. This requires care in formulating the approximating problem because it, too, must satisfy the compatibility condition (1.7) We shall address these issues in Chapter 8.

We note that problems of the third kind formally include the Dirichlet and Neumann problem. The examples considered later on are simply assumed (on physical grounds) to be well posed. Uniqueness, however, is easy to show with the maximum principle, provided

$$\alpha_1 \alpha_2 > 0.$$

As stated above, in this case we have a classical solution. Indeed, if u_1 and u_2 are two classical solutions of the Robin problem, then

$$w = u_1 - u_2$$

satisfies

$$\Delta w = 0, \qquad \vec{x} \in D$$

$$\alpha_1 \frac{\partial w}{\partial n} + \alpha_2 w = 0, \qquad \vec{x} \in \partial D.$$

From the maximum principle we know that w must assume its maximum and minimum on ∂D. Suppose that w has a positive maximum at $\vec{x}_0 \in \partial D$; then the boundary condition implies that

$$\frac{\partial w}{\partial n}(\vec{x}_0) = -\frac{\alpha_2}{\alpha_1} w(\vec{x}_0) < 0$$

so that w has a strictly positive directional derivative along the inward unit normal $-\vec{n}$. This contradicts that $w(\vec{x}_0)$ is a maximum of w on D. Hence w cannot have a positive maximum on $\bar{D}$. An analogous argument rules out a negative minimum so that $w \equiv 0$ is the only possibility.

Finally, let us illustrate the danger of imposing the wrong kind of boundary conditions on Laplace's equation.

For any positive integer k let us set

$$\epsilon_k = \frac{1}{(2k+1)\pi}$$

and consider the problem

$$\Delta u = 0, \qquad (x, y) \in (0, 1) \times (0, 1)$$

$$u = 0 \quad \text{on } x = 0, 1$$

and

$$u(x,0) = 0 \qquad \text{for } x \in (0,1)$$

$$u_y(x,0) = \epsilon_k \sin \frac{x}{\epsilon_k} \qquad \text{for } x \in (0,1).$$

We verify that

$$u_k(x,y) = \epsilon_k^2 \sin \frac{x}{\epsilon_k} \sinh \frac{y}{\epsilon_k}$$

is a solution of this problem. By inspection we see that

$$|u_{ky}(x,0)| \le \epsilon_k \to 0 \qquad \text{as } k \to \infty$$

while

$$\left| u\left(\frac{1}{2},1\right) \right| = \epsilon_k^2 \sinh \frac{1}{\epsilon_k} \to \infty \qquad \text{as } k \to \infty.$$

Hence the boundary data tend to zero uniformly while the solution blows up. Thus the problem cannot be well posed.

In general it is very dangerous to impose simultaneously Dirichlet *and* Neumann data (called Cauchy data) on the solution of an elliptic problem on a portion of ∂D even if the application does furnish such data. The resulting problem, even if formally solvable, tends to have an unstable solution.

1.4 The heat equation

The best known example of a parabolic equation is the M-dimensional heat equation

$$\mathcal{L}u \equiv \Delta u - u_t = F(\vec{x},t), \qquad \vec{x} \in D, \quad t > t_0 \qquad (1.8)$$

defined for the $(M+1)$-dimensional variable $(\vec{x},t) = (x_1,\ldots,x_M,t)$. It is customary to use t for the $(M+1)$st component because usually (but not always) time is a natural independent variable in the derivation of (1.8). For example, (1.8) is the mathematical model for the (scaled) temperature u in a homogeneous body D which changes through conduction in space and time. F represents a heat source when $F \le 0$ and a sink when $F > 0$. In this application the equation follows from the principle of conservation of energy and Fourier's law of heat conduction (see, e.g., [7]). However, quite diverse applications lead to (1.8) and to generalizations which formally look like (1.1). A parabolic equation like (1.1) with variable coefficients and additional terms is usually called the diffusion equation. We shall consider here the heat equation (1.8) because the qualitative behavior of its solution is generally a good guide to the behavior of the solution of a general diffusion equation.

We observe that a steady-state solution of (1.8) with a time-independent source term is simply the solution of Poisson's equation (1.2). Hence it is consistent to impose on (1.8) the same types of boundary conditions on ∂D discussed in Section 1.3 for Poisson's equation. Thus we speak of a Dirichlet, Neumann, or Robin problem (also known as a reflection problem) for (1.8). In addition, the application will usually provide an initial condition

$$u(\vec{x}, t_0) = u_0(\vec{x}), \qquad \vec{x} \in \bar{D}$$

at some time t_0 (henceforth set to $t_0 = 0$).

It is possible in applications that the domain D for the spacial variable $\vec{x}$ changes with time. However, separation of variables will require a time-independent domain. Hence D will be a fixed open set in $\mathbb{R}_M$ with boundary ∂D. The general formulation of an initial/boundary value problem for (1.8) is

$$\mathcal{L}u = F(\vec{x}, t), \qquad \vec{x} \in D, \quad t > 0 \tag{1.9a}$$

with initial condition

$$u(\vec{x}, 0) = u_0(\vec{x}), \qquad \vec{x} \in \bar{D} \tag{1.9b}$$

and boundary condition

$$\alpha_1 \frac{\partial u}{\partial n} + \alpha_2 u = g(\vec{x}, t), \qquad \vec{x} \in \partial D, \quad t > 0 \tag{1.9c}$$

for $\alpha_1, \alpha_2 \geq 0$ and $\alpha_1 + \alpha_2 > 0$. Let us define the set

$$Q_T = \{(\vec{x}, t) : \vec{x} \in D, \quad 0 < t \leq T\},$$

and the so-called parabolic boundary

$$\partial Q_T = \{(x, 0) : x \in \bar{D}\} \cup \{(\vec{x}, t) : \vec{x} \in \partial D, \quad 0 < t \leq T\},$$

where $T > 0$ is an arbitrary but fixed final time. Then a solution of (1.9) has to satisfy (in some sense) (1.9a) in Q_T and the boundary and initial conditions on ∂Q_T.

A classical solution of (1.9) is a function which is smooth in Q_T, which is continuous on $Q_T \cup \partial Q_T$ (together with its spacial derivatives if $\alpha_1 \neq 0$) and which satisfies the given data at every point of ∂Q_T.

It is common for diffusion problems that the initial and boundary conditions are not continuous at all points of the parabolic boundary. In this case u cannot be continuous on ∂Q_T and one again has to accept suitably defined weak solutions which only are required to solve the diffusion equation and initial/boundary conditions in an integral equation sense.

First and foremost in the discussion of well posedness for the heat equation is the question of existence of a solution. For the one-dimensional heat equation one can sometimes exhibit and analyze a solution in terms of an exponential integral — see the discussion of (1.13), (1.14) below — but in general this question is resolved with fairly abstract classical and functional analysis. As in the case of Poisson's equation it is possible to split the problem by writing

$$u = u_1 + u_2$$

where

$$\mathcal{L}u_1 = 0, \qquad (\vec{x}, t) \in Q_T$$

$$\alpha_1 \frac{\partial u_1}{\partial n} + \alpha_2 u_1 = g(\vec{x}, t), \qquad \vec{x} \in \partial D, \quad 0 < t \le T$$

$$u_1(\vec{x}, 0) = 0, \qquad \vec{x} \in \bar{D}$$

and

$$\mathcal{L}u_2 = F(\vec{x}, t), \qquad (\vec{x}, t) \in Q_T$$

$$\alpha_1 \frac{\partial u_2}{\partial n} + \alpha_2 u_2 = 0, \qquad \vec{x} \in \partial D, \quad 0 < t \le T$$

$$u_2(\vec{x}, 0) = u_0(\vec{x}), \qquad \vec{x} \in \bar{D}$$

and employ special techniques to establish the existence of u_1 and u_2. In particular, the problem for u_2 in an abstract sense has a lot in common with an n-dimensional first order system

$$\frac{du}{dt} - A(t)u = G(t), \qquad u(0) = u_0$$

and can be analyzed within the framework of (abstract) ordinary differential equations. We refer to the mathematics literature, notably [14], for an extensive discussion of classical and weak solutions of boundary value problems for linear and nonlinear diffusion equations. In general, it is safe to assume that if the boundary and initial data are continuous on the parabolic boundary, then all three types of initial/boundary value problems for the heat equation on a reasonable domain have unique classical solutions. Note that for the Neumann problem the heat equation does not require a compatibility condition linking the source F and the flux g.

If the data are discontinuous only at $t = 0$ (such as instantaneously heating an object at $t > 0$ on ∂D above its initial temperature), then the solution will be discontinuous at $t = 0$ but be differentiable for $t > 0$. It is useful to visualize such problems as the limit of problems with continuous but rapidly changing data near $t = 0$.

Continuous dependence of classical solutions for initial/boundary value problems on the data, and the uniqueness of the solution can be established with generalizations of the maximum principle discussed above for Poisson's equation. For example, we have the following analogues of Theorems 1.1 and 1.2.

Theorem 1.5 *The maximum principle*
Let u be a smooth solution of

$$\mathcal{L}u = F, \qquad (\vec{x}, t) \in Q_T.$$

If $F > 0$, then u cannot have a maximum in Q_T.

Proof. If u had a maximum at some point $(\vec{x}_0, t_0) \in Q_T$, i.e., $\vec{x}_0$ lies in the open set D, then necessarily

$$u_{x_i x_i}(\vec{x}_0, t_0) \le 0 \text{ and } u_t(\vec{x}_0, t_0) = 0 \text{ if } t_0 < T \text{ or } u_t(\vec{x}_0, t_0) \ge 0 \text{ if } t_0 = T.$$

In either case we could not satisfy

$$\mathcal{L}u(\vec{x}_0, t_0) = F(\vec{x}_0, t_0) > 0.$$

Arguments analogous to those applied above to Poisson's equation establish that if D is bounded and $F \geq 0$, then u must assume its maximum on ∂Q_T and if $F \leq 0$, then u must assume its minimum on ∂Q_T.

Continuous dependence of the solution of the Dirichlet problem for the heat equation on the data with respect to the sup norm is now defined as before. There is continuous dependence if for any $\epsilon > 0$, there exists a $\delta > 0$ such that

$$\|u\| \leq \epsilon$$

whenever $\|F\| + \max\{\|g\|, \|u_0\|\} \leq \delta$. Here

$$\|u\| = \max_{\bar{Q}_T} |u(x, t)|,$$

$$\|F\| = \sup_{Q_T} |F(x, t)|, \quad \|g\| = \max_{\substack{\partial \bar{D} \\ 0 \leq t \leq T}} |g(x, t)|, \quad \|u_0\| = \max_{\bar{D}} |u_0(x)|.$$

Theorem 1.6 *The solution of the Dirichlet problem for* (1.8) *depends continuously on the data F, g, and u_0.*

Proof. We again assume that the kth coordinate x_k satisfies $a \leq x_k \leq b$ for all $x \in D$. For arbitrary $\epsilon > 0$ define

$$\phi(\vec{x}) = (\|F\| + \epsilon)\frac{(x_k - a)^2}{2}.$$

Then

$$\mathcal{L}\left[\phi(\vec{x}) \pm u(\vec{x}, t)\right] \geq \epsilon > 0$$

and by Theorem 1.5 $[\phi \pm u]$ is bounded by its maximum on ∂Q_T. Hence

$$\pm u(\vec{x}, t) \leq \phi(\vec{x}) + u(\vec{x}, t) \leq (\|F\| + \epsilon) + \max\{\|g\|, \|u_0\|\}.$$

Continuous dependence now follows exactly as in the proof of Theorem 1.1 and we have the following analogue of estimate (1.5):

$$\|u\| \leq \|F\| \frac{(b - a)^2}{2} + \max\{\|g\|, \|u_0\|\}. \tag{1.10}$$

Similarly, we can conclude that the solution of the Dirichlet problem for the heat equation is unique and that the solution of the heat equation with $F \equiv 0$ (the homogeneous heat equation) must take on its maximum and minimum on ∂Q_T.

In addition to boundary value problems we also can consider a pure initial value problem for the heat equation

$$\mathcal{L}u \equiv \Delta u - u_t = 0, \qquad \vec{x} \in \mathbb{R}_M, \qquad t > 0$$

$$u(\vec{x}, 0) = u_0(\vec{x}).$$

It can be verified that the problem is solved by the formula

$$u(\vec{x}, t) = \int_{\mathbb{R}_M} s(\vec{x} - \vec{y}, t) u_0(\vec{y}) d\vec{y} \qquad (1.11)$$

where

$$s(\vec{x}, t) = \frac{1}{(4\pi t)^{M/2}} e^{-\frac{\langle \vec{x}, \vec{x} \rangle}{4t}}.$$

Here $\langle \vec{x}, \vec{x} \rangle$ denotes the dot product for vectors in $\mathbb{R}_M$. $s(\vec{x}, t)$ is known as the fundamental solution of the heat equation. A simple calculation shows that s is infinitely differentiable with respect to each component x_i and t for $t > 0$ and that

$$\Delta s(\vec{x}, t) - s_t(\vec{x}, t) = 0 \qquad \text{for } t > 0.$$

It follows from (1.11) that $u(\vec{x}, t)$ is infinitely differentiable with respect to all variables for $t > 0$ provided only that the resulting integrals remain defined and bounded. In particular, if u_0 is a bounded piecewise continuous function defined on $\mathbb{R}_M$, then the solution to the initial value problem exists and is infinitely differentiable for all x_i and all $t > 0$. It is harder to show that $u(\vec{x}, t)$ is continuous at $(\vec{x}_0, 0)$ at all points $\vec{x}_0$ where u_0 is continuous and that $u(\vec{x}, t)$ assumes the initial value $u_0(\vec{x}_0)$ as $(\vec{x}, t) \to (\vec{x}_0, 0)$. We refer to [5] for a proof of these results. Note that for discontinuous u_0 the expression (1.11) is only a weak solution because $u(\vec{x}, t)$ is not continuous at $t = 0$.

We see from (1.11) that if for any $\epsilon > 0$

$$u_0(\vec{x}) = \begin{cases} 1 & \text{for } \|\vec{x}\| \le \epsilon \\ 0 & \text{otherwise,} \end{cases}$$

then $u(\vec{x}, t) > 0$ for $t > 0$ at all $\vec{x} \in \mathbb{R}_M$. In other words, the initial condition spreads throughout space infinitely fast. This property is a consequence of the mathematical model and contradicts the observation that heat does not flow infinitely fast. But in fact, the change in the solution (1.11) at $\|\vec{x}\| \gg \epsilon$ remains unmeasurably small for a certain time interval before a detectable heat wave arrives so that defacto the wave speed is finite. We shall examine this issue at length in Example 6.3 where the speed of an isotherm is found numerically.

The setting of diffusion in all of $\mathbb{R}_M$ would seem to preclude the application of (1.11) to practical problems such as heat flow in a slab or bar. But (1.11) is not as restrictive as it might appear. This is easily demonstrated if $M = 1$. Suppose that u_0 is odd with respect to a given point x_0, i.e., $u_0(x_0 + x) = -u_0(x_0 - x)$; then with the obvious changes of variables

$$u(x_0 + x, t) = \int_{-\infty}^{\infty} s(x_0 + x - y, t) u_0(y) dy = \int_{-\infty}^{\infty} s(x - z, t) u_0(z + x_0) dz$$

$$= -\int_{-\infty}^{\infty} s(x - z, t) u_0(x_0 - z) dz = \int_{\infty}^{-\infty} s(x - x_0 + y) u_0(y) dy$$

$$= -\int_{-\infty}^{\infty} s(x_0 - x - y) u_0(y) dy = -u(x_0 - x, t)$$

we see that $u(x,t)$ is odd in x with respect to x_0 for all t. Since $u(x,t)$ is smooth for $t > 0$, this implies that $u(x_0,t) = 0$. Furthermore, if u_0 is periodic with period ω, then a similar change of variables techniques establishes that $u(x,t)$ is periodic in x with period ω. Hence if for an integer n

$$u_0(x) = \sin \frac{n\pi x}{L},$$

then u_0 is odd with respect to $x_0 = 0$ and $x_0 = L$ and the corresponding solution $u_n(x,t)$ given by (1.11) satisfies

$$u_n(0,t) = u_n(L,t) = 0 \qquad \text{for } t \geq 0.$$

It follows by superposition that if

$$u_0(x) = \sum_{n=1}^{N} \hat{\alpha}_n \sin \frac{n\pi x}{L} \tag{1.12}$$

for scalars $\{\hat{\alpha}_n\}$, then $u_N(x,t)$ given by (1.11) is the unique classical solution of the initial/boundary value problem

$$\mathcal{L}u \equiv u_{xx} - u_t = 0 \tag{1.13}$$
$$u(0,t) = u(L,t) = 0$$
$$u(x,0) = u_0(x).$$

Given any smooth function u_0 defined on $[0,L]$ it can be approximated by a finite sum (1.12) as discussed in Chapters 2 and 4. The corresponding solution $u_N(x,t)$ will be an approximate solution of problem (1.13). It can be shown by direct integration of (1.11) that this approximate solution is in fact identical to that obtained in Chapter 6 with our separation of variables approach. Note that if $u_0(0) \neq 0$ or $u_0(L) \neq 0$, then (1.13) does not have a classical solution and the maximum principle cannot be used to analyze the error $u(x,t) - u_N(x,t)$. Now continuous dependence in a mean square sense must be employed to examine the error. A simple version of the required arguments is given in Section 6.2 where error bounds for the separation of variables solution for the one-dimensional heat equation are considered. Of course, homogeneous boundary conditions are not realistic, but as we show time and again throughout the text, nonhomogeneous data can be made homogeneous at the expense of adding a source term to the heat equation. Thus instead of (1.13) one might have the problem

$$\mathcal{L}u \equiv u_{xx} - u_t = F(x,t) \tag{1.14}$$
$$u(0,t) = u(L,t) = 0$$
$$u(x,0) = 0.$$

We now verify by direct differentiation that if $U(x,t,\tau)$ is a solution of

$$\mathcal{L}U \equiv U_{xx}(x,t,\tau) - U_t(x,t,\tau) = 0$$

$$U(0, t, \tau) = U(L, t, \tau) = 0$$

$$U(x, \tau, \tau) = -F(x, \tau),$$

where τ is a nonnegative parameter, then

$$u(x, t) = \int_0^t U(x, t, \tau) d\tau \tag{1.15}$$

solves (1.14). The solution method leading to (1.15) is known as Duhamel's principle and can be interpreted as the superposition of solutions to the heat equation when the source is turned on only over a differential time interval dt centered at τ. The solution $U(x, t, \tau)$ is given by (1.11) as

$$U(x, t, \tau) = -\int_{-\infty}^{\infty} s(x - y, t - \tau) F(y, \tau) dy.$$

If $F(x, \tau)$ can be approximated by a sum of sinusoidal functions of period $2L$, then one can carry out all integrations analytically and obtain an approximate solution of (1.14). It again is identical with that found in Chapter 6. Note that the sum of the solutions of (1.13) and (1.14) solves the inhomogeneous heat equation with nonzero initial condition.

Similar results can be derived for an even initial function u_0 which allows the treatment of flux data at $x = 0$ or $x = L$. But certainly, this text promotes the view that the approach presented in Chapter 6 provides an easier and more general method for solving such boundary value problem than Duhamel's principle because the integration of (1.11) for a sinusoidal input and of (1.15) is replaced by an elementary differential equations approach.

Let us conclude our discussion of parabolic problems with a quick look at the so-called backward heat equation

$$\mathcal{L}u \equiv \Delta u + u_t = 0.$$

$$u(\vec{x}, 0) = u_0(\vec{x}).$$

Suppose we wish to find $u(\vec{x}, T)$ for $T > 0$. If we set $\tau = T - t$ and $w(\vec{x}, \tau) = u(\vec{x}, T - \tau)$, then the problem is equivalent to finding $w(\vec{x}, 0)$ of the problem

$$\mathcal{L}w \equiv \Delta w - w_\tau = 0$$

$$w(\vec{x}, T) = u_0(\vec{x}).$$

In a thermal setting this implies that from knowledge of the temperature at some future time T we wish to find the temperature today. Intuition tells us that if T is large and u_0 is near a steady-state temperature, then all initial temperatures $w(\vec{x}, 0)$ will decay to near u_0. In other words, small changes in u_0 could be consistent with large changes in $w(\vec{x}, 0)$, suggesting that the problem is not well posed when the heat equation is integrated backward in time (or the backward heat equation is integrated forward in time).

A well-known example is furnished by the function

$$u(x,t) = \epsilon e^{t/\epsilon^2} \sin(x/\epsilon), \qquad \epsilon > 0,$$

which solves the backward heat equation.

We see that

$$\lim_{\epsilon \to 0} u(x,0) = 0$$

but

$$\lim_{\epsilon \to 0} u(\pi\epsilon/2, t) = \infty \quad \text{for any } t > 0,$$

so there is no continuity with respect to the initial condition.

In general, any mathematical model leading to the backward heat equation which is to be solved forward in time will need to be treated very carefully. The comments at the end of Example 6.4 provide a further illustration of the difficulty of discovering the past from the present. Of course, if the backward heat equation is to be solved backward in time, as in the case of the celebrated Black–Scholes equation for financial options, then a time reversal will yield the usual well posed forward problem (see Example 6.9).

1.5   The wave equation

The third dominant equation of mathematical physics is the so-called wave equation

$$\mathcal{L}u \equiv \Delta u - \frac{1}{c^2} u_{tt} = 0, \qquad \vec{x} \in D \subset \mathbb{R}_M, \quad t \in (-\infty, \infty). \tag{1.16}$$

The equation is usually associated with oscillatory phenomena and shows markedly different properties compared to Poisson's and the heat equation. Equation (1.16) is an example of a hyperbolic equation. Here the $(M+1) \times (M+1)$ matrix $\mathcal{A}$ has the form

$$\mathcal{A} = \begin{pmatrix} I_M & 1 \\ 0 & -1 \end{pmatrix}$$

where I_M is the M-dimensional identity matrix.

It is easy to show that (1.16) allows wave-like solutions. For example, let f be an arbitrary twice continuously differentiable function of a scalar variable y. Let $\vec{n}$ be a (Euclidean) unit vector in $\mathbb{R}_M$ and define

$$y = \langle \vec{n}, \vec{x} \rangle - ct$$

where $\langle \vec{n}, \vec{x} \rangle$ is the dot product of $\vec{n}$ and $\vec{x}$. Then differentiation shows that

$$u(\vec{x}, t) = f(\langle \vec{n}, \vec{x} \rangle - ct)$$

is a solution of (1.16). Suppose that $c, t > 0$, then the set $\{\vec{x} : \langle \vec{n}, \vec{x} \rangle - ct = \text{constant}\}$ is a plane in $\mathbb{R}_M$ traveling in the direction of $\vec{n}$ with speed

c, and $f(\langle \vec{n}, \vec{x} \rangle - ct)$ describes a wave with constant value on this plane. For example, if

$$f(y) = e^{iy},$$

then

$$f(\langle \vec{n}, \vec{x} \rangle - ct) = e^{i[\langle \vec{n}, \vec{x} \rangle - ct]}$$

is known as a plane wave. Similarly, $g(\langle \vec{n}, \vec{x} \rangle + ct)$ describes a wave traveling in the direction of $-\vec{n}$ with speed c.

Our aim is to discuss again what constitutes well posed problems for (1.16). We begin by exhibiting a solution which is somewhat analogous to the solution of the one-dimensional heat equation discussed at the end of Section 1.4.

If we set $x = \langle \vec{n}, \vec{x} \rangle$, then the solutions $f(x - ct)$ and $g(x + ct)$ of (1.16) solve the one-dimensional wave equation

$$\mathcal{L}u \equiv u_{xx} - \frac{1}{c^2} u_{tt} = 0 \qquad (1.17)$$

which, for example, describes the motion of a vibrating uniform string. Here $u(x, t)$ is the vertical displacement of the string from its equilibrium position.

We show next that any smooth solution of (1.17) must be of the form

$$u(x, t) = f(x - ct) + g(x + ct),$$

i.e., the superposition of a right and left traveling wave. This observation follows if we introduce new variables

$$\xi = x - ct$$

$$\eta = x + ct$$

and express the wave equation in the new variables. The chain rule shows that

$$(\mathcal{L}u)(x, t) = 2u_{\xi\eta}(\xi, \eta) = 0$$

so that by direct integration

$$u(\xi, \eta) = f(\xi) + g(\eta) = f(x - ct) + g(x + ct)$$

for arbitrary continuously differentiable functions f and g.

For a vibrating string it is natural to impose an initial displacement and velocity of the string so that (1.17) is augmented with the initial conditions

$$u(x, 0) = u_0(x) \qquad (1.18)$$
$$u_t(x, 0) = u_1(x)$$

where u_0 and u_1 are given functions. Equations (1.18) constitute Cauchy data.

Let us suppose first that these functions are smooth and given on $(-\infty, \infty)$. Then we can construct a solution of (1.17), (1.18). We set

$$u(x, t) = f(x - ct) + g(x + ct)$$

and determine f and g so that u satisfies the initial conditions. Hence we need

$$u_0(x) = f(x) + g(x)$$

$$u_1(x) = -cf'(x) + cg'(x).$$

Integration of the last equation leads to

$$\frac{1}{c} \int_{x_0}^{x} u_1(s) ds = -f(x) + g(x) + K$$

where x_0 is some arbitrary but fixed point in $(-\infty, \infty)$ and $K = f(x_0) - g(x_0)$. When we solve algebraically for $f(x)$ and $g(x)$, we obtain

$$f(x) = \frac{1}{2} \left[u_0(x) - \frac{1}{c} \int_{x_0}^{x} u_1(s) ds + K \right]$$

$$g(x) = \frac{1}{2} \left[u_0(x) + \frac{1}{c} \int_{x_0}^{x} u_1(s) ds - K \right].$$

Hence

$$u(x,t) = \frac{1}{2} [u_0(x - ct) + u_0(x + ct)] + \frac{1}{2c} \left[-\int_{x_0}^{x-ct} u_1(s) ds + \int_{x_0}^{x+ct} u_1(s) ds \right]$$

which simplifies to

$$u(x,t) = \frac{1}{2} [u_0(x - ct) + u_0(x + ct)] + \frac{1}{2c} \int_{x-ct}^{x+ct} u_1(s) ds. \qquad (1.19)$$

The expression (1.19) is known as d'Alembert's solution for the initial value problem of the one-dimensional wave equation. If u_0 is twice continuously differentiable and u_1 is once continuously differentiable on $(-\infty, \infty)$, then the d'Alembert solution is a classical solution of the initial value problem for all finite t and x. Moreover, it is unique because u has to be the superposition of two traveling waves and the d'Alembert construction determines f and g uniquely. Moreover, if we set

$$\|u_k\| = \sup_{(-\infty, \infty)} |u_k(x)|, \quad k = 0, 1,$$

then

$$|u(x,t)| \le \|u_0\| + t\|u_1\|$$

which implies continuous dependence for all $t \le T$ where T is an arbitrary but fixed time. Hence the initial value problem (1.17), (1.18) is well posed.

We observe from (1.19) that the value of $u(x_0, t_0)$ at a given point (x_0, t_0) depends only on the initial value u_0 at $x_0 - ct_0$ and $x_0 + ct_0$ and on the initial value u_1 over the interval $[x_0 - ct_0, x_0 + ct_0]$. Thus, if

$$u_0(x) = u_1(x) = 0 \quad \text{for } |x| \ge \epsilon$$

and $x_0 > \epsilon$, then regardless of the form of the data on the set $|x| < \epsilon$ we have

$$u(x_0, t) = \begin{cases} 0 & \text{for } t < \frac{x_0 - \epsilon}{c} \\ \frac{1}{2c} \int_{-\epsilon}^{\epsilon} u_1(s) ds & \text{for } t > \frac{x_0 + \epsilon}{c} . \end{cases}$$

Hence these initial conditions travel with speed c to the point x_0 but in general $u(x_0, t)$ will not decay to zero as $t \to \infty$. This is a peculiarity of the M-dimensional wave equation for $M = 1$ and all even M [5]. If u_0 and u_1 do not have the required derivatives but (1.19) remains well defined, then (1.19) represents a weak solution of (1.17), (1.18). We shall comment on this aspect when discussing a plucked string in Example 7.1.

As in the case of the heat kernel solution we can exploit symmetry properties of the initial conditions to solve certain initial/boundary value problems for the one-dimensional wave equation with d'Alembert's solution. For example, suppose that u_0 and u_1 are smooth and odd with respect to the point x_0; then the d'Alembert solution is odd with respect to x_0.

To see this suppose that

$$u(x, t) = \int_{x-ct}^{x+ct} u_1(s) ds$$

and that u_1 is odd with respect to the point x_0, i.e., $u_1(x_0 + x) = -u_1(x_0 - x)$. Then with $y = x_0 - s$ and $r = x_0 + y$ we obtain

$$u(x_0 + x, t) = \int_{x_0 + x - ct}^{x_0 + x + ct} u_1(s) ds = - \int_{-x+ct}^{-x-ct} u_1(x_0 - y) dy$$

$$= \int_{-x+ct}^{-x-ct} u_1(x_0 + y) dy = \int_{x_0 - x + ct}^{x_0 - x - ct} u_1(r) dr = -u(x_0 - x, t).$$

Since by hypothesis

$$u_0(x_0 + x - ct) + u_0(x_0 + x + ct) = -u_0(x_0 - x + ct) - u_0(x_0 - x - ct),$$

we conclude that the d'Alembert solution is odd with respect to the point x_0 and hence equal to zero at x_0 for all t. It follows that the boundary value problem

$$\mathcal{L}u \equiv u_{xx} - \frac{1}{c^2} u_{tt} = 0$$

$$u(0, t) = u(L, t) = 0$$

$$u(x, 0) = u_0(x) = \sum_{n=1}^{N} \hat{\alpha}_n \sin \frac{n\pi x}{L}$$

$$u_t(x, 0) = u_1(x) = \sum_{n=1}^{N} \hat{\beta}_n \sin \frac{n\pi}{L}$$

for constant $\{\hat{\alpha}_n\}$ and $\{\hat{\beta}_n\}$ is solved by the d'Alembert solution (1.19) because u_0 and u_1 are defined on $(-\infty, \infty)$ and odd with respect to $x = 0$ and $x = L$.

As we remarked in Section 1.4, inhomogeneous boundary conditions can often be made homogeneous at the expense of adding a source term to the differential equation. This leads to problems of the type

$$\mathcal{L}u \equiv u_{xx} - \frac{1}{c_2} u_{tt} = F(x,t)$$

$$u(0,t) = u(L,t) = 0$$

$$u(x,0) = u_t(x,0) = 0.$$

Now a Duhamel superposition principle can be applied. It is straightforward to show that the function

$$u(x,t) = \int_0^t U(x,t,\tau)d\tau$$

solves our problem whenever $U(x,t,\tau)$ is the solution of

$$\mathcal{L}U \equiv U_{xx} - \frac{1}{c^2} U_{tt} = 0$$

$$U(0,t,\tau) = U(L,t,\tau) = 0$$

$$U(x,\tau,\tau) = 0$$

$$U_t(x,\tau,\tau) = -c^2 F(x,\tau)$$

where τ is a parameter. It follows that if F is of the form

$$F_N(x,\tau) = \sum_{n=1}^{N} \gamma_n(\tau) \sin \frac{n\pi x}{L},$$

then the problem has the d'Alembert solution

$$U(x,t,\tau) = -\frac{c}{2} \int_{x-c(t-\tau)}^{x+c(t-\tau)} F_N(s,\tau)ds.$$

All integrations can be carried out analytically and the resultant solution $u_N(x,t)$ can be shown to be identical to the separation of variables solution found in Chapter 7 when an arbitrary source term F is approximated by a trigonometric sum F_N.

Let us now turn to the general initial/boundary value problem of the form

$$\Delta u - u_{tt} = F(\vec{x},t), \qquad \vec{x} \in D, \quad t > 0 \qquad (1.20)$$

$$u(\vec{x},0) = u_0(\vec{x}) \qquad \vec{x} \in D$$

$$u_t(\vec{x},0) = u_1(\vec{x})$$

$$\alpha_1 \frac{\partial u}{\partial n} + \alpha_2 u = g(\vec{x}, t), \qquad \vec{x} \in \partial D, \quad t > 0$$

where D is a given domain in $\mathbb{R}_M$. For convenience we have set $c = 1$ which can always be achieved by scaling time. We point out that if $D \equiv \mathbb{R}_M$ and we have a pure initial value problem, then it again is possible to give a formula for $u(x, t)$ analogous to the d'Alembert solution of the one-dimensional problem (see [5, Chapter 2]), but for a true initial/boundary value problem the existence of a solution will generally be based on abstract theory. (We note in this context that if $g \equiv 0$ on ∂D, then, again in a very general sense, the problem has a lot in common with the ordinary differential equation

$$\frac{d^2 u}{dt^2} - A(t)u = G(t), \quad u(0) = u_0, \quad u'(0) = u_1.)$$

We shall henceforth assume that the existence theory of [5] applies so that we can concentrate on uniqueness of the solution and on its continuous dependence on the data of the problem.

Uniqueness and continuous dependence follow from a so-called energy estimate. If u is a smooth solution of (1.20) with $g \equiv 0$ on ∂D, then

$$u_t \mathcal{L} u \equiv u_t \Delta u - u_t u_{tt} = u_t F(\vec{x}, t)$$

so that

$$\int_D [\nabla \cdot u_t \nabla u - \nabla u_t \cdot \nabla u - u_t u_{tt}] \, dx = \int_D u_t F \, d\vec{x}.$$

We now apply the divergence theorem and obtain

$$\frac{d}{dt} \int_D \frac{1}{2} [\nabla u \cdot \nabla u + u_t^2] \, d\vec{x} - \oint_{\partial D} u_t \frac{\partial u}{\partial n} \, d\vec{s} = -\int_D u_t F \, d\vec{x}. \qquad (1.21)$$

For a smooth solution the boundary data

$$\alpha_1 \frac{\partial u}{\partial n} + \alpha_2 u = 0 \quad \text{with } \alpha_1 + \alpha_2 = 1$$

imply

$$\alpha_1 \frac{\partial u_t}{\partial n} + \alpha_2 u_t = 0$$

so that

$$(\alpha_1 + \alpha_2) u_t \frac{\partial u}{\partial n} = -\alpha_2 u_t u - \alpha_1 \frac{\partial u}{\partial n} \left(\frac{\partial u}{\partial n} \right)_t.$$

Since $|u_t F| \leq \frac{u_t^2}{2} + \frac{F^2}{2}$, we obtain from (1.21) the estimate

$$\frac{d}{dt} E(t) \leq E(t) + \frac{\|F(\cdot, t)\|^2}{2} \qquad (1.22)$$

where

$$E(t) = \frac{1}{2} \int_D [\nabla u \cdot \nabla u + u_t^2] \, d\vec{x} + \frac{1}{2} \oint_{\partial D} \left[\alpha_2 u^2 + \alpha_1 \left(\frac{\partial u}{\partial n} \right)^2 \right] d\vec{s}$$

and

$$\|F(\cdot,t)\| = \left(\int_D F^2(\vec{x},t)d\vec{x} \right)^{1/2}.$$

The inequality (1.22) can be written as

$$E'(t) = E(t) + \frac{\|F(\cdot,t)\|^2}{2} - g(t)$$

where g is some unknown nonnegative function. This differential equation has the analytic solution

$$E(t) = E(0)e^t + \int_0^t e^{t-r} \left[\frac{\|F(\cdot,r)\|^2}{2} - g(r) \right] dr$$

from which we obtain the so-called Gronwall inequality

$$E(t) \le E(0)e^t + \int_0^t e^{t-r} \left[\frac{\|F(\cdot,r)\|^2}{2} \right] dr \tag{1.23}$$

where

$$E(0) = \frac{1}{2} \left[\int_D \nabla u_0 \cdot \nabla u_0 + u_1^2 \right] d\vec{x} + \frac{1}{2} \oint_{\partial D} \left[\alpha_2 u_0^2 + \alpha_1 \left(\frac{\partial u_0}{\partial n} \right)^2 \right] d\vec{s}$$

is known from the initial data.

We have the following two immediate consequences of (1.23).

Theorem 1.7 *The solution of the initial/boundary value problem* (1.20) *is unique.*

Proof. The difference w of two solutions satisfies (1.20) with

$$F = g = u_0 = u_1 = 0.$$

This implies that $E(0) = 0$ so that (1.23) assures that $E(t) = 0$ for all t. Then by Schwarz's inequality (see Theorem 2.4)

$$|w(x,t)|^2 = \left| \int_0^t w_s(x,s)ds \right|^2 \le t \int_0^t w_s^2(x,s)ds,$$

from which follows that

$$\|w(\cdot,t)\|^2 \equiv \int_D w(x,t)^2 dx \le 2t \int_0^t E(s)ds = 0.$$

Hence $w = 0$ in the mean square sense for all t which implies that a classical solution is identically zero.

Theorem 1.7 assures that the separation of variables solution constructed in Section 7.1 is the only solution of the approximating problem. Similar arguments are used in Section 7.2 to show for a vibrating string that this approximate solution converges to the analytic solution of the original problem as the approximations of the data are refined.

Chapter 2

Basic Approximation Theory

This chapter will review the abstract ideas of approximation that will be used in the sequel. Let X be a linear space (sometimes called a vector space) over the field S of real or complex numbers. The elements of X are called vectors and those of S are called scalars. The vector spaces appearing in this book are R_n and C_n with elements $\vec{x} = (x_1, \ldots, x_n)$, $\vec{y}$, etc., or spaces of real- or complex-valued functions f, g, etc. defined on a real interval or, more generally, a subset of Euclidean n-space. In all spaces $\vec{0}$ denotes the zero vector. The scalars of S are denoted by α, β, or a, b, etc.

We now recall a few definitions from linear algebra which are central in our discussion of the approximation of functions.

Definition Given a collection $\mathcal{C} = \{\varphi_1, \varphi_2, \ldots, \varphi_N\}$ of vectors in a linear space X, then

$$M \equiv \text{span}\{\varphi_1, \ldots, \varphi_N\}$$

is the set of all linear combinations $\{\alpha_1\varphi_1 + \cdots + \alpha_N\varphi_N\}$ of the elements of $\mathcal{C}$.

Note that M is a subspace of X because it is closed under vector addition and scalar multiplication.

Definition A collection $\mathcal{C} = \{\varphi_1, \varphi_2, \ldots, \varphi_N\}$ of vectors in a linear space is linearly independent if

$$\sum_{j=1}^{N} \alpha_j\varphi_j = \vec{0} \quad \text{only if} \quad \alpha_1 = \alpha_2 = \cdots = \alpha_N = 0.$$

A sequence of vectors $\{\varphi_n\}$ is linearly independent if any finite collection $\mathcal{C}$ drawn from the sequence is linearly independent.

The definition implies that in a finite set of linearly independent vectors no one element can be expressed as a linear combination of the remaining elements.

Definition Let $\mathcal{C} = \{\varphi_1, \ldots, \varphi_N\}$ be a collection of N linearly independent vectors of X with the property that every element of X is a linear combination of the $\{\varphi_j\}$ (i.e., span$\{\varphi_1, \ldots, \varphi_N\} = X$). Then the set $\{\varphi_j\}$ is a basis of X, and X has dimension N.

The theorems of linear algebra assure that every basis of an N-dimensional vector space consists of N elements, but not every vector space has a finite-dimensional basis. Spaces containing sets of countably many linearly independent vectors are called infinite dimensional.

2.1   Norms and inner products

Basic to the idea of approximation is the concept of a distance between vectors f and an approximation g, or the "size" of the vector $f - g$. This leads us to the the idea of a norm for assessing the size of vectors.

Definition Let X be a vector space. A norm on X is a real-valued function $F : X \to R$ such that for every $f, g \in X$ and every scalar $\alpha \in S$, it is true that

i) $F(f) \neq 0$ if $f \neq \vec{0}$;

ii) $F(\alpha f) = |\alpha| \, F(f)$;

iii) $F(f + g) \leq F(f) + F(g)$.

Proposition 2.1 $F(f) \geq 0$.

Proof. From ii) with $\alpha = 0$, we know that $F(\vec{0}) = 0$. Thus

$$0 = F(f + (-f)) \leq F(f) + F(-f) = 2F(f).$$

In other words, $F(f) \geq 0$.

The value of the norm function F is almost always written as $\|f\|$. A vector space X together with a norm on X is called a normed linear space. The inequality iii) is commonly called the triangle inequality.

The concept of a norm is an abstraction of the usual length of a vector in Euclidean three-space and provides a measure of the "distance" $\|f - g\|$ between two vectors $f, g \in X$. Thus in the space R_3 of triples $\vec{x} = (x_1, x_2, x_3)$ of real numbers with the customary definitions of addition and scalar multiplication, the function

$$\|\vec{x}\| = \sqrt{x_1^2 + x_2^2 + x_3^2}$$

is a norm (see Example 2.6a). The distance induced by this norm is the everyday Euclidean distance.

In a linear space X, our approximation problem will be to find a member f_M of a given subspace $M \subset X$ that is closest to a given vector f in the sense that $\|f - f_M\| \leq \|f - m\|$ for all $m \in M$. We shall be concerned only with finite-dimensional subspaces M.

We know from elementary geometry that in ordinary Euclidean three-space, the closest point on a line or a plane containing the origin to a given point $\vec{x}$ in the space is the perpendicular projection of $\vec{x}$ onto the line or plane. This is the idea that we shall abstract to our general setting. For this, we need to extend the notion of the "dot," or scalar, product.

Definition An inner product on a vector space X is a scalar-valued function $G : X \times X \rightarrow S$ on ordered pairs of elements of X such that

i) $G(f, f) \geq 0$ and $G(f, f) = 0$ if and only if $f = \vec{0}$;

ii) $G(f, g) = \overline{G(g, f)}$ [$\bar{\beta}$ denotes the complex conjugate of β];

iii) $G(\alpha f, g) = \alpha G(f, g)$; and

iv) $G(f + g, h) = G(f, h) + G(g, h)$.

We shall usually denote $G(f, g)$ by $\langle f, g \rangle$. A vector space together with an inner product defined on it is called an inner product space.

The next proposition is easy to verify.

Proposition 2.2 *An inner product has the following properties:*

i) $\langle f, \alpha g \rangle = \bar{\alpha} \langle f, g \rangle$,

ii) $\langle f, g + h \rangle = \langle f, g \rangle + \langle f, h \rangle$,

iii) $\left\langle \vec{0}, g \right\rangle = 0$.

Definition Two vectors f and g in an inner product space are said to be orthogonal if $\langle f, g \rangle = 0$.

Example 2.3 a) In real Euclidean n-space R_n, the usual dot product

$$\langle \vec{x}, \vec{y} \rangle = \vec{x} \cdot \vec{y} = \sum_{j=1}^{n} x_j y_j,$$

where $\vec{x} = (x_1, x_2, \ldots, x_n)$ and $\vec{y} = (y_1, y_2, \ldots, y_n)$ is an inner product.

b) In R_2, for $\vec{x} = (x_1, x_2)$ and $\vec{y} = (y_1, y_2)$ define $\langle \vec{x}, \vec{y} \rangle$ by $\langle \vec{x}, \vec{y} \rangle = A\vec{x} \cdot \vec{y}$, where A is the matrix

$$A = \begin{pmatrix} 2 & 1 \\ 1 & 2 \end{pmatrix}$$

and $\vec{u} \cdot \vec{v}$ is the usual dot product of $\vec{u}$ and $\vec{v}$. Then $\langle \vec{x}, \vec{y} \rangle$ is an inner product. First consider

$$\langle \vec{x}, \vec{x} \rangle = A\vec{x} \cdot \vec{x} = (2x_1 + x_2, x_1 + 2x_2) \cdot (x_1, x_2) = 2\left(x_1^2 + x_2 x_1 + x_2^2\right)$$
$$= 2\left[\left(x_1 + \frac{x_2}{2}\right)^2 + \frac{3}{4}x_2^2\right].$$

It is clear that $\langle \vec{x}, \vec{x} \rangle \geq 0$ and $\langle \vec{x}, \vec{x} \rangle = 0$ if and only if $\vec{x} = (0, 0)$.

To see that $\langle \vec{x}, \vec{y} \rangle = \langle \vec{y}, \vec{x} \rangle$, simply compute both inner products. The remaining two properties are evident.

c) On the space of all continuous functions (real- or complex-valued) defined on the reals having period $2L$, it is easy to verify that

$$\langle f, g \rangle = \int_{-L}^{L} f(t)\overline{g(t)}dt$$

is an inner product.

Theorem 2.4 *In an inner product space,* $|\langle f, g \rangle| \leq \sqrt{\langle f, f \rangle}\sqrt{\langle g, g \rangle}$.

Proof. If $\langle f, g \rangle = 0$, then the proposition is obviously true, so assume $\langle f, g \rangle \neq 0$. Let α be a complex number. Then

$$\langle f + \alpha g, f + \alpha g \rangle \geq 0.$$

Now

$$\langle f + \alpha g, f + \alpha g \rangle = \langle f, f \rangle + \overline{\alpha}\langle f, g \rangle + \alpha\langle g, f \rangle + |\alpha|^2 \langle g, g \rangle.$$

Next, let $\alpha = t\langle f, g \rangle$, where t is any real number. Then

$$\langle f + \alpha g, f + \alpha g \rangle = \langle f, f \rangle + \overline{\alpha}\langle f, g \rangle + \alpha\langle g, f \rangle + \langle g, g \rangle$$
$$= \langle f, f \rangle + 2t\,|\langle f, g \rangle|^2 + t^2\,|\langle f, g \rangle|^2 \langle g, g \rangle \geq 0.$$

This expression is quadratic in t and so the fact that it is never negative means that

$$4\,|\langle f, g \rangle|^4 - 4\,|\langle f, g \rangle|^2 \langle f, f \rangle \langle g, g \rangle \leq 0.$$

In other words,

$$|\langle f, g \rangle|^2 \leq \langle f, f \rangle \langle g, g \rangle,$$

which completes the proof.

The inequality

$$|\langle f, g \rangle| < \sqrt{\langle f, f \rangle}\,\sqrt{\langle g, g \rangle}$$

is known as Schwarz's inequality.

Corollary 2.5 *Suppose X is an inner product space. Then the function F defined by $F(f) = \sqrt{\langle f, f \rangle}$ is a norm on X.*

Proof. The proofs that $F(f) \geq 0$ and $F(\alpha f) = |\alpha|\,F(f)$ are simple and omitted. We prove the triangle inequality.

$$F(f + g)^2 = \langle f + g, f + g \rangle = \langle f, f \rangle + \langle f, g \rangle + \langle g, f \rangle + \langle g, g \rangle$$
$$= F(f)^2 + 2\mathrm{Re}(\langle f, g \rangle) + F(g)^2 \leq F(f)^2 + 2\,|\langle f, g \rangle| + F(g)^2$$
$$\leq F(f)^2 + 2F(f)F(g) + F(g)^2 = (F(f) + F(g))^2.$$

Hence $F(f + g) \leq F(f) + F(g)$, and we see that $\sqrt{\langle f, f \rangle}$ is indeed a norm on X.

In an inner product space, the norm $\|f\| = \sqrt{\langle f, f \rangle}$ is called the norm induced by the inner product $\langle \cdot, \cdot \rangle$.

Example 2.6 a) Let R_n be real Euclidean n-space endowed with the usual inner product $\langle \vec{x}, \vec{y} \rangle = \sum_{j=1}^{n} x_j y_j$ (cf. Example 2.3a). Then the norm induced by this inner product is the usual Euclidean norm

$$\|\vec{x}\| = \sqrt{x_1^2 + x_2^2 + \cdots + x_n^2}.$$

b) Let X be the space of all complex valued continuous functions defined of the interval $[a, b]$ (cf. Example 2.3c). Then

$$\langle f, g \rangle = \int_a^b f(t) \overline{g(t)} dt$$

is an inner product on X and

$$\|f\| = \sqrt{\int_a^b |f(t)|^2 \, dt}$$

is the norm induced by this inner product. Very closely related to this example is the root mean square of a function on the interval $[a, b]$

$$rms(f) = \|f\| = \sqrt{\frac{\int_a^b |f(t)|^2 \, dt}{b - a}}.$$

2.2 Projection and best approximation

In Euclidean space, given a line or a plane M through the origin and a vector $\vec{v}$, the vector in M closest to $\vec{v}$ is the vector $\vec{y}$ such that $\vec{v} - \vec{y}$ is perpendicular to every vector in M. This vector $\vec{y}$ is the projection of $\vec{v}$ onto M.

We shall see that the idea of a projection generalizes to the abstract setting of inner product spaces.

Definition Suppose M is a subspace of an inner product space X and suppose $f \in X$. The orthogonal projection of f onto M is a vector $Pf \in M$ such that

$$r = f - Pf$$

is orthogonal to every $m \in M$.

In case M has finite dimension, the existence of an orthogonal projection onto M is easy to establish. Suppose $\{\varphi_1, \varphi_2, \ldots, \varphi_N\}$ is a basis for the subspace M. The projection is a member of M and so

$$Pf = \sum_{j=1}^{N} \alpha_j \varphi_j,$$

and we need only to find the coordinates α_j. First, observe that a vector is orthogonal to every element of M if and only if it is orthogonal to each of the basis elements φ_i. Hence we want

$$\langle f - P_M f, \varphi_i \rangle = \langle f, \varphi_i \rangle - \left\langle \sum_{j=1}^{N} \alpha_j \varphi_j, \varphi_i \right\rangle = 0 \quad \text{for } i = 1, 2, \ldots, N.$$

Since $\left\langle \sum_{j=1}^{N} \alpha_j \varphi_j, \varphi_i \right\rangle = \sum_{j=1}^{N} \alpha_j \langle \varphi_j, \varphi_i \rangle$, we have

$$\sum_{j=1}^{N} \alpha_j \langle \varphi_j, \varphi_i \rangle = \langle f, \varphi_i \rangle, \qquad i = 1, 2, \ldots, N.$$

In matrix-vector form

$$\mathcal{A}\vec{\alpha} = \vec{b},$$

where

$$\mathcal{A} = (a_{ij}) \text{ with } a_{ij} = \langle \varphi_j, \varphi_i \rangle \text{ and } \vec{b} = (b_i) \text{ with } b_i = \langle f, \varphi_i \rangle.$$

Proposition 2.7 *The coefficient matrix $\mathcal{A} = (\langle \varphi_j, \varphi_i \rangle)$ is nonsingular.*

Proof. Suppose $\mathcal{A}\vec{\beta} = \vec{0}$ for some vector $\vec{\beta} = (\beta_1, \beta_2, \ldots, \beta_N)$. Then

$$\left\| \sum_{j=1}^{N} \beta_j \varphi_j \right\|^2 = \left\langle \sum_{j=1}^{N} \beta_j \varphi_j, \sum_{i=1}^{N} \beta_i \varphi_i \right\rangle = \sum_{i=1}^{N} \bar{\beta}_i \sum_{j=1}^{N} \langle \varphi_j, \varphi_i \rangle \beta_j = \mathcal{A}\vec{\beta} \cdot \vec{\beta} = 0.$$

But the collection $\{\varphi_1, \varphi_2, \cdots, \varphi_N\}$ is independent and so it must be true that $\beta_1 = \beta_2 = \cdots = \beta_N = 0$. Hence $\mathcal{A}$ is nonsingular.

We have thus shown that the system $\mathcal{A}\vec{\alpha} = \vec{b}$ has a unique solution for the coordinates of the orthogonal projection of f onto M. In other words, the orthogonal projection of f onto M exists and is unique. We can thus safely speak of Pf as being *the* orthogonal projection of f onto M with respect to the inner product $\langle \cdot, \cdot \rangle$.

Remark If the basis $\{\varphi_1, \varphi_2, \ldots, \varphi_N\}$ is orthogonal (i.e., $\langle \varphi_i, \varphi_j \rangle = 0$ for $i \neq j$), then A is a diagonal matrix, and the coordinates of Pf are easily found

$$Pf = \sum_{i=1}^{N} \frac{\langle f, \varphi_i \rangle}{\langle \varphi_i, \varphi_i \rangle} \varphi_i.$$

Next we see that the projection of f onto a subspace M is the member of M that best approximates f in the sense that of all elements m of M, it is one that makes $\|m - f\|$ the smallest.

Theorem 2.8 *Let M be a subspace of an inner product space X and suppose $f \in X$. If $m \in M$, then*
$$\|f - Pf\| \le \|f - m\|,$$
where Pf is the orthogonal projection of f onto M.

Proof. Observe that

$$
\begin{aligned}
\|f - m\|^2 &= \|f - Pf + Pf - m\|^2 \\
&= \langle (f - Pf) + (Pf - m), (f - Pf) + (Pf - m) \rangle \\
&= \langle f - Pf, f - Pf \rangle + \langle f - Pf, Pf - m \rangle \\
&\quad + \langle Pf - m, f - Pf \rangle + \langle Pf - m, Pf - m \rangle.
\end{aligned}
$$

Then

$$
\begin{aligned}
\|f - m\|^2 &= \langle f - Pf, f - Pf \rangle + \langle Pf - m, Pf - m \rangle \\
&= \|f - Pf\|^2 + \|Pf - m\|^2
\end{aligned}
$$

since $\langle f - Pf, Pf - m \rangle = 0 = \overline{\langle Pf - m, f - Pf \rangle} = 0$ because $Pf - m \in M$. Thus
$$\|f - Pf\|^2 = \|f - m\|^2 - \|Pf - m\|^2, \text{ or}$$
$$\|f - Pf\|^2 \le \|f - m\|^2.$$

Hence the orthogonal projection Pf of f onto M is a best approximation. We next show that any best approximation of f from the subspace M must be this unique projection of f onto M.

Theorem 2.9 *Let M be a subspace of an inner product space X and suppose $f \in X$. If $h \in M$ is such that $\|f - h\| \le \|f - m\|$ for all $m \in M$, then $h = Pf$, the orthogonal projection of f onto M.*

Proof. Let $m \in M$ be an arbitrary element of M. For any real t and any scalar α, the vector $h + t\alpha m$ is a member of M. Thus the function F defined by

$$F(t) = \|f - (h + t\alpha m)\|^2 = \langle f - h - t\alpha m, f - h - t\alpha m \rangle$$

has a minimum at $t = 0$. But $F(t)$ is simply a quadratic function of the real variable t

$$F(t) = \langle f - h, f - h \rangle - t \langle f - h, \alpha m \rangle - t \langle \alpha m, f - h \rangle + t^2 \langle \alpha m, \alpha m \rangle.$$

$F'(0) = 0$ means

$$
\begin{aligned}
\langle f - h, \alpha m \rangle + \langle \alpha m, f - h \rangle &= \langle f - h, \alpha m \rangle + \overline{\langle f - h, \alpha m \rangle} \\
&= 2\operatorname{Re} \langle f - h, \alpha m \rangle = 2\operatorname{Re} \left(\overline{\alpha} \langle f - h, m \rangle \right) = 0.
\end{aligned}
$$

Now simply choose the scalar $\alpha = \langle f - h, m \rangle$. Then

$$\mathrm{Re}\left(\overline{\alpha}\,\langle f - h, m \rangle\right) = |\langle f - h, m \rangle|^2 = 0.$$

Hence $\langle f - h, m \rangle = 0$, which shows that h must be the orthogonal projection of f onto M.

Example 2.10 In the plane R_2, let $\vec{x} = (3,5)$, $\vec{y} = (1,2)$, and let $M = \mathrm{span}\,\{\vec{x}\}$.

a) With the usual inner product, which induces the usual Euclidean length of vectors, the projection $P\vec{y}$ of $\vec{y}$ onto M is

$$P\vec{y} = \frac{\langle \vec{y}, \vec{x} \rangle}{\langle \vec{x}, \vec{x} \rangle}\,\vec{x} = \frac{1 \cdot 3 + 2 \cdot 5}{3^2 + 5^2}\,(3,5) = \left(\frac{39}{34}, \frac{65}{34}\right).$$

Then $\left(\frac{39}{34}, \frac{65}{34}\right)$ is the point on the line $5x - 3y = 0$ that is closest to $(1,2)$ and the vector $\vec{y} - P\vec{y} = \left(-\frac{5}{34}, \frac{3}{34}\right)$ is perpendicular to all vectors $\alpha\,(3,5) \in M$.

b) In the plane with the inner product of Example 2.3b, the projection $P\vec{y}$ of $\vec{y}$ onto the subspace M is

$$P\vec{y} = \frac{\langle \vec{y}, \vec{x} \rangle}{\langle \vec{x}, \vec{x} \rangle}\,\vec{x} = \frac{A\vec{y} \cdot \vec{x}}{A\vec{x} \cdot \vec{x}}\,\vec{x}, \quad \text{where } A = \left(\begin{array}{cc} 2 & 1 \\ 1 & 2 \end{array}\right).$$

Then $A\vec{y} = (4,5)$, $A\vec{x} = (11,13)$, and so

$$P\vec{y} = \frac{37}{98}\,(3,5).$$

c) In the space of all continuous real-valued functions on the interval $[-1,1]$ with the inner product

$$\langle f, g \rangle = \int_{-1}^{1} f(t)g(t)dt,$$

let M be the subspace consisting of all polynomials of degree ≤ 2. Let us find the orthogonal projection onto M of f defined by

$$f(x) = \begin{cases} 0 & x < 0 \\ x & 0 < x. \end{cases}$$

The obvious basis for M is $\{1, x, x^2\}$. Then

$$\langle 1, x \rangle = \langle x, 1 \rangle = \int_{-1}^{1} t\,dt = 0; \quad \langle 1, x^2 \rangle = \langle x^2, 1 \rangle = \int_{-1}^{1} t^2\,dt = \frac{2}{3}$$

$$\langle x, x^2 \rangle = \langle x^2, x \rangle = \int_{-1}^{1} t^3 dt = 0$$

$$\langle 1, 1 \rangle = 2; \quad \langle x, x \rangle = \frac{2}{3}; \quad \text{and } \langle x^2, x^2 \rangle = \int_{-1}^{1} t^4 dt = \frac{2}{5}.$$

The coefficient matrix A for the linear system to be solved for the coordinates of the projection Pf is

$$A = \begin{pmatrix} 2 & 0 & 2/3 \\ 0 & 2/3 & 0 \\ 2/3 & 0 & 2/5 \end{pmatrix}.$$

Next

$$\langle f, 1 \rangle = \int_0^1 t\, dt = \frac{1}{2}; \quad \langle f, x \rangle = \int_0^1 t^2\, dt = \frac{1}{3}; \quad \text{and } \langle f, x^2 \rangle = \int_0^1 t^3\, dt = \frac{1}{4}.$$

We now know that $Pf(x) = \alpha_1 + \alpha_2 x + \alpha_3 x^3$, where

$$\begin{pmatrix} 2 & 0 & 2/3 \\ 0 & 2/3 & 0 \\ 2/3 & 0 & 2/5 \end{pmatrix} \begin{pmatrix} \alpha_1 \\ \alpha_2 \\ \alpha_3 \end{pmatrix} = \begin{pmatrix} 1/2 \\ 1/3 \\ 1/4 \end{pmatrix}.$$

Thus $\alpha_1 = 3/32$, $\alpha_2 = 1/2$, and $\alpha_3 = 15/32$. The projection of f is then

$$Pf(x) = \frac{3}{32} + \frac{1}{2}x + \frac{15}{32}x^2.$$

We have found the "best" approximation to f in the sense that of all quadratic functions q, this is the one that gives the smallest value of

$$\|f - q\|^2 = \int_{-1}^1 (f(x) - q(x))^2 dx.$$

This is sometimes called the *least squares* approximation.

Pictures of both f and Pf on the same axes are shown in Fig. 2.1.

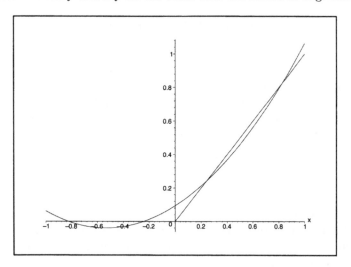

Figure 2.1: Least squares approximation of $f(x) = \max\{0, x\}$ on $[-1, 1]$ with a quadratic polynomial.

2.3   Important function spaces

It is easy to see that the collection of all continuous (real- or complex-valued) functions defined on an interval $[a, b]$ with the usual definition of addition is a linear space, traditionally denoted $C[a, b]$. It becomes an inner product space if we define $\langle f, g \rangle$ by

$$\langle f, g \rangle = \int_a^b f(x)\overline{g(x)}dx.$$

A so-called weight function w may also be introduced. If the function w is real valued and continuous on $[a, b]$ and such that $w(x) \geq 0$ for all x, and $w(x) = 0$ at a finite set of points, then it is easy to see that

$$\langle f, g \rangle = \int_a^b f(x)\overline{g(x)}w(x)dx.$$

is also an inner product for our space. The restriction on $w(x)$ guarantees that $\langle f, f \rangle > 0$ for $f \neq 0$.

Spaces of continuous functions are useful, but they are not sufficiently large for subsequent applications because they do not contain certain important types of functions — step functions, square waves, unbounded functions, etc. The spaces with which we shall be primarily concerned are the so-called L_2 spaces. Specifically, suppose D is a real interval, finite or infinite, and w is a weight function as defined above. Then $L_2(D, w)$ is the collection of all functions for which $|f|^2 w$ is integrable. It can be shown that this is indeed a vector space and that with the definition

$$\langle f, g \rangle = \int_D f(x)\overline{g(x)}w(x)dx,$$

we have almost an inner product space: "almost" because with this definition, it is possible to have $\langle f, f \rangle = \|f\|^2 = 0$ for a function f other than the zero function. For example, with $w(x) \equiv 1$, $D = [0, 1]$, and f given by $f(0) = 1$ and $f(x) = 0$ for all $x \neq 0$, we have $\|f\| = 0$. To ensure we have an inner product, we simply say that two functions f and g are "equal" if

$$\|f - g\|^2 = \int_D |f(x) - g(x)|^2 w(x)dx = 0.$$

Here by "integrable" we mean integrable in the sense of Lebesgue, but the reader unfamiliar with this concept need not be concerned. In the applications in the sequel, the integrals encountered will all be the usual Riemann integrals of elementary calculus.

Note. In case the weight function $w(x) \equiv 1$, we abbreviate $L_2(D, w)$ with $L_2(D)$.

The essence of this discussion remains the same when we consider real- or complex-valued functions of several variables; i.e., when the domain D is a subset of R_n.

Given $f \in L_2(D, w)$ then its orthogonal projection onto a finite-dimensional subspace is just a least squares approximation (as in Example 2.10c), but with respect to the weight function w. The weight function is sometimes introduced to improve the fit of the approximation in the region where w is large, but in this text its role is usually to make the basis spanning M an orthogonal basis as we shall see in the next chapter.

Example 2.11 a) Let $D = [0, 1]$. Then in the space $L_2(D)$, it is straightforward to see that the set $B = \{1, \cos \pi x, \cos 2\pi x, \cos 3\pi x\}$ is orthogonal. We shall find the projection Pf of the step function f given by

$$f(x) = -1 + 2H(x - .5),$$

onto the subspace $M = \text{span } B$.

Here H is the Heaviside function defined by

$$H(x) = \begin{cases} 0, & x < 0 \\ 1, & x > 0. \end{cases}$$

Since B is orthogonal, we obtain

$$
\begin{aligned}
Pf &= \frac{\langle f, 1 \rangle}{\langle 1, 1 \rangle} 1 + \frac{\langle f, \cos \pi x \rangle}{\langle \cos \pi x, \cos \pi x \rangle} \cos \pi x + \frac{\langle f, \cos 2\pi x \rangle}{\langle \cos 2\pi x, \cos 2\pi x \rangle} \cos 2\pi x \\
&\quad + \frac{\langle f, \cos 3\pi x \rangle}{\langle \cos 3\pi x, \cos \pi x \rangle} \cos 3\pi x \\
&= \frac{-2/\pi}{1/2} \cos \pi x + \frac{2/3\pi}{1/2} \cos 3\pi x \\
&= -\frac{4}{\pi} \cos \pi x + \frac{4}{3\pi} \cos 3\pi x.
\end{aligned}
$$

b) Let us again project f onto span B, but in the space $L_2(D, w)$, where $w(x) = x(1 - x)$. The inner product is now given by

$$\langle g, h \rangle = \int_0^1 g(x)h(x)x(1 - x)dx,$$

and the collection B is no longer orthogonal. The projection Pf in this case is

$$Pf = \alpha_1 + \alpha_2 \cos \pi x + \alpha_3 \cos 2\pi x + \alpha_4 \cos 3\pi x,$$

where

$$\begin{pmatrix} \langle 1,1 \rangle & \langle \cos\pi x,1 \rangle & \langle \cos 2\pi x,1 \rangle & \langle \cos 3\pi x,1 \rangle \\ \langle 1,\cos\pi x \rangle & \langle \cos\pi x,\cos\pi x \rangle & \langle \cos 2\pi x,\cos\pi x \rangle & \langle \cos 3\pi x,\cos\pi x \rangle \\ \langle 1,\cos 2\pi x \rangle & \langle \cos\pi x,\cos 2\pi x \rangle & \langle \cos 2\pi x,\cos 2\pi x \rangle & \langle \cos 3\pi x,\cos 2\pi x \rangle \\ \langle 1,\cos 3\pi x \rangle & \langle \cos\pi x,\cos 3\pi x \rangle & \langle \cos 2\pi x,\cos 3\pi x \rangle & \langle \cos 3\pi x,\cos 3\pi x \rangle \end{pmatrix}$$

$$\times \begin{pmatrix} \alpha_1 \\ \alpha_2 \\ \alpha_3 \\ \alpha_4 \end{pmatrix} = \begin{pmatrix} \langle f,1 \rangle \\ \langle f,\cos\pi x \rangle \\ \langle f,\cos 2\pi x \rangle \\ \langle f,\cos 3\pi x \rangle \end{pmatrix}.$$

After some computation, we have

$$\begin{pmatrix} 0.16667 & 0 & 0.05066 & 0 \\ 0 & 0.05800 & 0 & -0.03166 \\ 0.05066 & 0 & 0.07700 & 0 \\ 0 & -0.03166 & 0 & 0.08052 \end{pmatrix} \begin{pmatrix} \alpha_1 \\ \alpha_2 \\ \alpha_3 \\ \alpha_4 \end{pmatrix} = \begin{pmatrix} 0 \\ -0.08552 \\ 0 \\ 0.08034 \end{pmatrix},$$

and $\alpha_1 = \alpha_3 = 0$, $\alpha_2 = -1.1840$, and $\alpha_4 = 0.53224$.

Our projection Pf thus becomes

$$Pf = -1.1840 \cos\pi x + 0.53224 \cos 3\pi x.$$

A picture of f, the projection from a) and the projection just found are shown in Fig. 2.2.

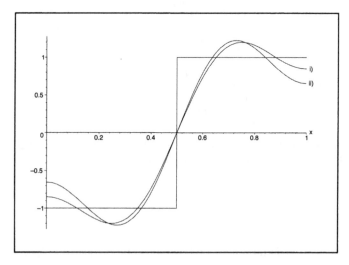

Figure 2.2: Orthogonal projections of $f(x) = -1 + 2H(x - .5)$. i) in $L_2(0,1)$ and ii) in $L_2(0,1,x(1-x))$.

Finally, we note that on occasion for a function of several variables we shall consider a subset of them as fixed parameters and compute projections with respect to the remaining variables. For example, let $F(x,t)$ be a function defined on a set $D \times T$, where D and T are intervals. Suppose that $F(x,t) \in L_2(D,w)$ for each $t \in T$. Let $M = \text{span} \{\varphi_1, \varphi_2, \ldots, \varphi_N\} \subset L_2(D,w)$. Then we write

$$PF(x,t) = \sum_{j=1}^{N} \alpha_j(t)\varphi_j(x).$$

If F depends smoothly on t, then it follows from the computation of the $\alpha_j(t)$ that these coefficients likewise will depend smoothly on the parameter t.

In approximation theory the function $F(\cdot, t)$ is interpreted as an abstract function defined on $[0,T]$ with values in $L_2(D,w)$. The approximation $PF(\cdot, t)$ then is the associated abstract function with values in the finite-dimensional subspace M. However, we shall be content to consider t simply as a parameter.

Example 2.12 Let $F(x,t) = \cos(xt)$ for all $(x,t) \in D \times R$, where D is the interval $[0,1]$ and R is the entire real line. Then for each real t, it is clear that $F(x,t) \in L_2(D)$. We know that $\{1, \cos \pi x, \cos 2\pi x, \cos 3\pi x\}$ is orthogonal. Let us find the projection of F onto $M = \text{span} \{1, \cos \pi x, \cos 2\pi x, \cos 3\pi x\}$.

$$\langle F, 1 \rangle = \int_0^1 F(x,t)dx = \int_0^1 \cos xt \, dx = \frac{\sin t}{t}$$

$$\langle F, \cos \pi x \rangle = \int_0^1 \cos xt \cos \pi x \, dx = \frac{t \sin t}{\pi^2 - t^2}$$

$$\langle F, \cos 2\pi x \rangle = \int_0^1 \cos xt \cos 2\pi x \, dx = \frac{t \sin t}{t^2 - 4\pi^2}$$

$$\langle F, \cos 3\pi x \rangle = \int_0^1 \cos xt \cos 3\pi x \, dx = \frac{t \sin t}{9\pi^2 - t^2}.$$

From Example 2.11a) we know that

$$\langle 1, 1 \rangle = 1, \quad \text{and}$$
$$\langle \cos \pi x, \cos \pi x \rangle = \langle \cos 2\pi x, \cos 2\pi x \rangle = \langle \cos 3\pi x, \cos 3\pi x \rangle = 1/2.$$

Thus

$$PF = \frac{\sin t}{t} + 2\frac{t \sin t}{\pi^2 - t^2} \cos \pi x + 2\frac{t \sin t}{t^2 - 4\pi^2} \cos 2\pi x + 2\frac{t \sin t}{9\pi^2 - t^2} \cos 3\pi x.$$

This appears at first glance to contradict the statement that the coefficients should be differentiable functions of t, but notice that each term of the sum has a limit at the zero of the denominator. If, as we usually do, define them to be equal to this limit at the value of t at which the expression is undefined, computation will show the resulting functions to be differentiable. Look for instance at $\alpha_2(t)$, the coefficient of $\cos \pi x$. Then

$$\lim_{t \to \pi} 2 \frac{t \sin t}{\pi^2 - t^2} = 2 \lim_{t \to \pi} \frac{t}{\pi + t} \lim_{t \to \pi} \frac{\sin(t - \pi)}{t - \pi} = 1.$$

The same limit is obtained as $t \to -\pi$ and so

$$\alpha_2(t) = \begin{cases} 2 \frac{t \sin t}{\pi^2 - t^2}, & t \neq \pm \pi \\ 1, & t = \pm \pi \end{cases}$$

is continuous at $t = \pm \pi$. In fact, since $\sin t / t$ is infinitely differentiable at $t = 0$ one can readily show that $\alpha_2(t)$ has derivatives of all orders there.

We make the observation that even though $F(x, t)$ may be well behaved it often is not possible to find $PF(x, t)$ explicitly as a function of t. A projection of $F(x, t)$ onto span$\{\varphi_1, \ldots, \varphi_N\}$ requires the coefficient

$$\alpha_j(t) = \frac{\langle F(x, t), \varphi_j(x) \rangle}{\langle \varphi_j, \varphi_j \rangle}$$

which is available only if $\langle F(x, t), \varphi_j \rangle$ can be evaluated analytically or numerically for arbitrary t. For a function like

$$F(x, t) = \frac{1}{f(t) + g(x)}$$

this is generally not possible. On the other hand, for a given value t^* the integral

$$\langle F(x, t^*), \varphi_j \rangle$$

will always be assumed known, if not analytically, then at least numerically.

We shall conclude with a result which has the same flavor as Theorems 2.8 and 2.9 and which is used in Example 7.7.

Theorem 2.13 *Let X be an inner product space. Let $\{\varphi_1, \ldots, \varphi_N\}$ be a set of N linearly independent elements of X. Then the "smallest" solution (i.e., the minimum norm solution) of the N linear equations*

$$\langle \varphi_i, u \rangle = b_i, \qquad i = 1, \ldots, N$$

belongs to

$$M = \text{span}\{\varphi_1, \ldots, \varphi_N\}.$$

Proof. We show first that the equations have a unique solution u_N in M. If we substitute

$$u_N = \sum_{j=1}^{N} \alpha_j \varphi_j$$

into the linear equations, we find that the $\{\alpha_j\}$ must solve the the matrix equation

$$\mathcal{A}\vec{\alpha} = \vec{b}$$

where

$$\mathcal{A}_{ij} = \langle \varphi_i, \varphi_j \rangle.$$

We know from Proposition 2.7 that $\mathcal{A}$ is nonsingular so that there is a unique solution $u_N \in M$.

Suppose there is another solution $u \in X$. Then

$$\langle u, u \rangle = \langle u - u_N + u_N, u - u_N + u_N \rangle$$
$$= \langle u - u_N, u - u_N \rangle + \langle u_N, u_N \rangle + 2\mathrm{Re}\langle u_N, u - u_N \rangle.$$

Since u and u_N satisfy the same equations, it follows that

$$\langle u_N, u - u_N \rangle = \sum_{n=1}^{N} \alpha_j \langle \varphi_j, u - u_N \rangle = 0.$$

Hence $\langle u_N, u_N \rangle < \langle u, u \rangle$ for any other solution u.

Exercises

2.1) Find the equation of the form $ax + by + cz = d$ spanned by the vectors $\vec{x} = (1, 2, 0)$ and $\vec{y} = (0, 1, 1)$.

2.2) Prove or disprove: the functions $\{\sin t, \cos t, \sin(t+5)\}$ are linearly independent.

2.3) Let C_n be complex Euclidean space; i.e., the space of all n-tuples $\vec{x} = (x_1, x_2, \ldots, x_n)$ of complex numbers with usual definitions of vector addition and scalar multiplication. Show that each of the following defines a norm on C_n:
   i) $\|\vec{x}\| = \max\{|x_j| : j = 1, 2, \ldots, n\}$,
   ii) $\|\vec{x}\| = \sum_{j=1}^{n} |x_j|$.

2.4) Let X be the vector space of all continuous complex-valued functions f on the interval $[0, 1]$. Show that each of the following defines a norm on X:
   i) $\|f\| = \sup\{|f(x)| : x \in [0, 1]\}$,
   ii) $\|f\| = \int_0^1 |f(t)| \, dt$.

2.5) Prove Proposition 2.2.

2.6) On complex Euclidean space C_n, define $\langle \vec{x}, \vec{y} \rangle = \sum\limits_{j=1}^{n} x_j \overline{y_j}$, where $\vec{x} = (x_1, x_2, \ldots, x_n)$ and $\vec{y} = (y_1, y_2, \ldots, y_n)$. Verify that this defines an inner product on C_n.

2.7) Verify Example 2.3c.

2.8) Show that in an inner product space X, the vector $\vec{0}$ is orthogonal to every vector in X, and it is the only element of X having this property.

2.9) Let X be an inner product space and let $v \in X$. Prove that the set S of all $x \in X$ orthogonal to v is a subspace of X. (This subspace is called the orthogonal complement of v.)

2.10) Let $\vec{v} = (1, 2) \in R_2$.
   i) Find the orthogonal complement of $\vec{v}$ with respect to the usual dot product.
   ii) Find the orthogonal complement of $\vec{v}$ with respect to the inner product described in Example 2.3b.

2.11) Let M be a subspace of an inner product space X and let $Q \subset X$ be given by
$$Q = \{f \in X : \langle f, m \rangle = 0 \text{ for every } m \in M\}.$$

   i) Show that Q is a subspace of X.
   ii) Show that $X = \{q + m : q \in Q \text{ and } m \in M\}$.
   [The subspace Q is called the orthogonal complement of M; thus the orthogonal complement of a vector v defined in Exercise 2.9 is in this sense the orthogonal complement of the subspace spanned by v.]

2.12) In R_2 let $\vec{x} = (2, 3)$. Find
   i) $\|\vec{x}\|$ where the norm is the one induced by the usual dot product (Example 2.3a);
   ii) $\|\vec{x}\|$ where the norm is the one induced by the inner product described in Example 2.3b.

2.13) In R_2, show that
$$\|\vec{x}\| = \|(x_1, x_2)\| = \max\{|x_1|, |x_2|\}$$

defines a norm. For $\vec{x} = (1, 2)$ and $\vec{y} = (2, 1)$, find $\|\vec{x}\|$, $\|\vec{y}\|$, $\|\vec{x} + \vec{y}\|$, and $\|\vec{x} - \vec{y}\|$.

2.14) Find the root mean square of $f(x) = \sin \frac{2\pi x}{b}$ on the interval $[0, b]$.

2.15) Suppose X is an inner product space. Show that for $\|\vec{x}\| = \sqrt{\langle \vec{x}, \vec{x} \rangle}$, it is true that
$$\|\vec{x} + \vec{y}\|^2 + \|\vec{x} - \vec{y}\|^2 = 2\|\vec{x}\|^2 + 2\|\vec{y}\|^2.$$

2.16) Is every norm induced by some inner product?

2.17) In an inner product space, show that if $\langle f, g \rangle = 0$, then $\|f + g\|^2 = \|f\|^2 + \|g\|^2$. This is called the Pythagorean theorem. Why?

2.18) Let $0 = t_0 < t_1 < \cdots < t_N = L$ define a partition on $[0, L]$ with $t_{n+1} - t_n = \Delta t$. Define the mapping T from $C[0, L]$ into $\mathbb{R}_{N+1}$
$$Tf = (f(t_0), f(t_1), \ldots, f(t_N)).$$

i) Show that T is a linear transformation.
ii) Show that T is not invertible.
iii) Find an inner product $\langle , \rangle$ on $\mathbb{R}_{N+1}$ such that
$$\int_0^L f(t)g(t)dt \cong \langle Tf, Tg \rangle$$

for small Δt.

2.19) In the Euclidean plane with the usual norm, given a point $\vec{x} = (x_1, x_2)$, find the point on the line $y = ax$ closest to $\vec{x}$ by projecting the point onto the line.

2.20) In Euclidean three-space with the usual norm, find the point in the plane $2x + y - 3z = 0$ that is closest to the point $(0, 0, 5)$.

2.21) Find the point on the line described by the vector function $\vec{F}(t) = (t + 1, 2t - 1, -t)$ that is closest to the point $(1, 4, -2)$.

2.22) Find the polynomial of degree ≤ 2 that is the best approximation to $f(x) = \cos \frac{\pi}{2} x$ in the sense that among all such polynomials q it minimizes
$$\int_{-1}^{1} \left(\cos \frac{\pi}{2} x - q(x) \right)^2 dx.$$

Sketch the graphs of your approximation and f on the same axes.

2.23) Suppose M is a finite-dimensional subspace of an inner product space, and suppose $f \in M$. What is the orthogonal projection of f onto M?

2.24) Let X be an inner product space. For $f \in X$, let M be the orthogonal complement of f. What is the orthogonal projection of f onto$M$?

2.25) Suppose X is an inner product space with an orthogonal basis $B = \{\varphi_1, \varphi_2, \ldots, \varphi_n\}$. Let $f = \sum_{j=1}^{n} c_j \varphi_j$, and let $C = \{\varphi_{i(1)}, \varphi_{i(2)}, \ldots, \varphi_{i(k)}\}$ be a subset of B. Find Pf, the orthogonal projection of f onto $M = $ span C.

2.26) In an inner product space X, let $B = \{\varphi_1, \varphi_2, \ldots, \varphi_n\}$ be a basis for the subspace M. Define a sequence of vectors $\gamma_1, \gamma_2, \ldots, \gamma_n$ as follows:

$\gamma_1 = \varphi_1$;

$\gamma_j = \varphi_j - P_{j-1}\varphi_j$, $j = 2, 3, \ldots n$, where $P_{j-1}\varphi_j$ is the orthogonal projection of φ_j onto span$\{\gamma_1, \gamma_2, \ldots, \gamma_{j-1}\}$.

Prove that $\tilde{B} = \{\gamma_1, \gamma_2, \ldots, \gamma_n\}$ is an orthogonal basis for M. [This recipe for finding an orthogonal basis is called the Gram–Schmidt process.]

2.27) Let X be the subspace of $L^2(R)$ consisting of all real-valued functions with period 2π with the inner product

$$\langle f, g \rangle = \int_{-\pi}^{\pi} f(t)g(t)dt.$$

i) Verify that the collection $C = \{\sin x, \sin 2x, \sin 3x\}$ is orthogonal.

ii) Let f be the periodic extension of

$$\widehat{f}(x) = \begin{cases} -1 & \text{for } -\pi < x \le 0 \\ 1 & \text{for } 0 < x \le \pi, \end{cases}$$

and find the projection of f onto the space spanned by the collection C.

iii) Sketch the graphs of f and the projection found in ii) on the same axes.

2.28) Given N continuous functions $\{f_i^0(t)\}$ and N distinct points $\{t_j\}$, show that if the $N \times N$ matrix A with entries

$$A_{ij} = f_i(t_j)$$

is nonsingular, then the functions are linearly independent. Use the example $f_1(t) = t(1 - t)$ and $f_2(t) = t^2(1 - t)$, and $t_1 = 0$ and $t_2 = 1$ to show that if the matrix

$$\{f_i(t_j)\}$$

is singular, then we cannot conclude anything about the linear dependence of $\{f_i(t)\}$.

2.29) Suppose $\{f_i(t)\}$ is a set of N functions, each of which is $N - 1$ times continuously differentiable on $(0, 2)$. Show that if the $N \times N$ matrix A with entries

$$A_{ij} = \left\{ f_j^{(i-1)}(t) \right\}$$

is nonsingular at some point $t^* \in (0, L)$, then the functions $\{f_i(t)\}$ are linearly independent. Use the example $f_1(t) = \max\{(1 - t)^3, 0\}$, $f_2(t) = \max\{0, (t - 1)^3\}$ to show that if A is singular, then we can conclude nothing about linear dependence. (Note: The determinant of the matrix A is known as the Wronskian of the functions $\{f_i(t)\}$.)

Chapter 3

Sturm–Liouville Problems

Many of the problems to be considered later will require an approximation of given functions in terms of eigenfunctions of an ordinary differential operator. A well-developed eigenvalue theory exists for so-called Sturm–Liouville differential operators, and we shall summarize the results important for the solution of partial differential equations later on. However, in many applications only very simple and readily solved eigenvalue problems arise which do not need the generality of the Sturm–Liouville theory. We shall consider such problems first.

3.1 Sturm–Liouville problems for $\phi'' = \mu\phi$

The simplest, but also constantly recurring, operator is

$$\mathcal{L}\phi \equiv \phi''(x) \qquad 0 < x < L$$

defined on the vector space $C^2(0, L)$ of twice continuously differentiable functions on the interval $(0, L)$, or on some subspace M of $C^2(0, L)$ determined by the boundary conditions to be imposed on ϕ. Henceforth M will denote the domain on which $\mathcal{L}$ is to be defined. In analogy to the matrix eigenvalue problem

$$A\phi = \lambda\phi$$

for an $n \times n$ matrix A we shall consider the following problem:

Find an eigenvalue μ and all eigenfunctions (= eigenvectors) $\phi(x) \in M$ which satisfy

$$\mathcal{L}\phi = \mu\phi \quad \text{for all } x \in (0, L). \tag{3.1}$$

As in the matrix case the eigenvalue may be zero, real, or complex, but the corresponding eigenfunction must not be the zero function. Note that if ϕ is an eigenvector, then $c\phi$ for $c \neq 0$ is also an eigenvector.

The domain on which $\mathcal{L}$ is defined has an enormous influence on the solvability of the eigenvalue problem. For example, if $M = C^2(0, L)$, then for any complex number μ the equation

$$\mathcal{L}\phi \equiv \phi'' = \mu\phi$$

has the two linearly independent solutions

$$\phi_1 = e^{\sqrt{\mu}x}$$

$$\phi_2 = e^{-\sqrt{\mu}x}$$

for $\mu \neq 0$ where $+\sqrt{\mu}$ and $-\sqrt{\mu}$ denote the two roots of $z^2 - \mu = 0$, and

$$\phi_1 = 1$$

$$\phi_2 = x$$

for $\mu = 0$. Hence any number μ is an eigenvalue and has two corresponding eigenfunctions. On the other hand, if $M = \{\phi \in C^2(0, L) : \phi(0) = \phi'(0) = 0\}$, then for any μ the only solution of

$$\phi'' = \mu\phi$$

$$\phi(0) = \phi'(0) = 0$$

is the zero solution. Hence there are no eigenvalues and eigenvectors in this case.

The subspaces M of $C^2(0, L)$ of interest for applications are defined by the so-called Sturm–Liouville boundary conditions

$$
\begin{array}{llll}
\text{i)} & \phi(0) = 0, & \phi(L) = 0 & \\
\text{ii)} & \phi'(0) = 0, & \phi(L) = 0 & \\
\text{iii)} & \phi(0) = 0, & \phi'(L) = 0 & \\
\text{iv)} & \phi'(0) = 0, & \phi'(L) = 0 & \\
\text{v)} & \alpha\phi'(0) = \phi(0), & \phi(L) = 0, & \alpha > 0 \\
\text{vi)} & \alpha\phi'(0) = \phi(0), & \phi'(L) = 0, & \alpha > 0 \\
\text{vii)} & \phi(0) = 0, & \beta\phi'(L) = -\phi(L), & \beta > 0 \\
\text{viii)} & \phi'(0) = 0, & \beta\phi'(L) = -\phi(L), & \beta > 0 \\
\text{ix)} & \alpha\phi'(0) = \phi(0), & \beta\phi'(L) = -\phi(L), & \alpha, \beta > 0 \\
\text{x)} & \phi(0) = \phi(L), & \phi'(0) = \phi'(L). &
\end{array}
\tag{3.2}
$$

For example, in case iii) M subspaces consist of all functions $\phi \in C^2(0, L)$ such that $\phi(0)$ and $\phi'(L)$ are defined and

$$\lim_{x \to 0} \phi(x) = \phi(0) = 0$$

$$\lim_{x \to L} \phi'(x) = \phi'(L) = 0.$$

We note that the first nine boundary conditions represent special cases of the general condition

$$\alpha_1\phi'(0) - \alpha_2\phi(0) = 0, \qquad (3.3)$$
$$\beta_1\phi'(L) + \beta_2\phi(L) = 0;$$

for real α_i, β_j such that

$$\alpha_1\alpha_2 \geq 0, \qquad \alpha_1^2 + \alpha_2^2 \neq 0;$$
$$\beta_1\beta_2 \geq 0, \qquad \beta_1^2 + \beta_2^2 \neq 0.$$

The boundary condition x) is associated with periodic functions defined on the line. In each case the subspace M will consist of those functions in $C^2(0, L)$ which are continuous or continuously differentiable at 0 and L and which satisfy the given boundary conditions.

For several of these boundary conditions we can give explicitly the eigenvalues and eigenfunctions. To see what is involved let us look at the simple case of

$$\phi'' = \mu\phi,$$
$$\phi(0) = \phi(L) = 0.$$

If $\mu = 0$, then $\phi(x) = c_1 + c_2 x$, and the boundary conditions require $c_1 = c_2 = 0$ so that $\phi(x) \equiv 0$; hence $\mu = 0$ is not an eigenvalue. For $\mu \neq 0$ the differential equation has again the general solution

$$\phi(x) = c_1 e^{\sqrt{\mu}x} + c_2 e^{-\sqrt{\mu}x}.$$

The two boundary conditions require

$$c_1 + c_2 = 0, \quad \text{and}$$

$$c_1 e^{\sqrt{\mu}L} + c_2 e^{-\sqrt{\mu}L} = 0;$$

or in matrix form

$$\begin{pmatrix} 1 & 1 \\ e^{\sqrt{\mu}L} & e^{-\sqrt{\mu}L} \end{pmatrix} \begin{pmatrix} c_1 \\ c_2 \end{pmatrix} = \begin{pmatrix} 0 \\ 0 \end{pmatrix}.$$

This system has a nontrivial solution $(c_1, c_2) = (1, -1)$ if and only if the determinant $f(\mu)$ of the coefficient matrix is zero. We need

$$f(\mu) \equiv e^{-\sqrt{\mu}L} - e^{\sqrt{\mu}L} = e^{-\sqrt{\mu}L}(1 - e^{2\sqrt{\mu}L}) = 0.$$

This can be the case only if $e^{2\sqrt{\mu}L} = 1$, i.e., if

$$2\sqrt{\mu}L = 2n\pi i$$

for a nonzero integer n. Hence there are countably many eigenvalues

$$\mu_n = - \left(\frac{n\pi}{L} \right)^2$$

with corresponding eigenfunction

$$\phi_n(x) = c \left(e^{i\frac{n\pi}{L}x} - e^{-i\frac{n\pi}{L}x} \right).$$

Since eigenfunctions are determined only up to a multiplicative constant, we can choose $c = \frac{1}{2i}$ so that

$$\phi_n(x) = \sin \frac{n\pi}{L} x.$$

The above calculation would have been simpler had we known a priori that the eigenvalue has to be real and nonnegative. In that case complex numbers and functions can be avoided as we shall see below. For the algebraic sign pattern of the coefficients in the boundary conditions of all of the above ten eigenvalue problems this property is easy to establish.

Theorem 3.1 *The eigenvalues of* (3.1) *for the boundary conditions* (3.2i–x) *are real and nonpositive.*

Proof. If $\{\mu, \phi(x)\}$ is an eigenvalue–eigenvector pair then

$$\int_0^L \phi''(x)\overline{\phi(x)}dx = \mu \int_0^L \phi(x)\overline{\phi(x)}dx.$$

Integration by parts shows that

$$\int_0^L \phi''(x)\overline{\phi(x)}dx = \phi'(x)\overline{\phi(x)} \Big|_0^L - \int_0^L \phi'(x)\overline{\phi'(x)}dx.$$

For each of the ten cases above the boundary terms either vanish or are real and nonpositive. For example, if $\beta_2 \neq 0$ in (3.3), then

$$\phi'(L)\overline{\phi(L)} = -\frac{\beta_1}{\beta_2} \phi'(L)\overline{\phi'(L)} \leq 0.$$

Hence

$$\mu \int_0^L |\phi(x)|^2 dx \leq \int_0^L |\phi'(x)|^2 dx \leq 0$$

which implies that μ is real and nonpositive.

Since the eigenvalue is real, it follows that the real and imaginary part of any complex-valued eigenfunction must satisfy the eigenvalue equation. Hence the eigenfunctions may be taken to be real so that the conjugation in the integrals can be dropped.

Theorem 3.2 *The eigenfunctions of* (3.1), (3.2) *corresponding to distinct eigen-values are orthogonal in* $L_2(0, L)$.

Proof. Let $\{\mu_m, \phi_m\}$ and $\{\mu_n, \phi_n\}$ be eigenvalues and eigenfunctions with $\mu_m \neq \mu_n$. Then

$$(\mu_m - \mu_n) \int_0^L \phi_m(x)\phi_n(x)dx = \int_0^L (\phi_n\phi_m'' - \phi_m\phi_n'')dx.$$

If we integrate by parts and use the boundary conditions, we see that the right-hand integral vanishes so that

$$\langle \phi_m, \phi_n \rangle \equiv \int_0^L \phi_m(x)\phi_n(x)dx = 0.$$

The computation of the eigenvalues and eigenvectors for the above cases is straightforward, at least in principle. Since $\mu \leq 0$, we shall write

$$\mu = -\lambda^2$$

and solve

$$\phi'' + \lambda^2\phi = 0.$$

We know that the general solution of this equation is

$$\phi(x) = c_1 + c_2 x, \qquad \lambda = 0 \tag{3.4}$$

$$\phi(x) = c_1 \cos \lambda x + c_2 \sin \lambda x, \qquad \lambda \neq 0. \tag{3.5}$$

We now have to determine from the boundary conditions for what values of λ we can find a nontrivial solution. To introduce the required computations let us look at the simple cases (3.2ii) and (3.2x) before considering the general case (3.3).

(3.2ii) $\phi'(0) = \phi(L) = 0$.

We see by inspection that $\lambda = 0$ is not admissible because there is no non-zero eigenfunction of the form (3.4). For $\lambda \neq 0$ the solution is given by (3.5). The boundary conditions lead to

$$\lambda c_2 = 0$$

$$c_1 \cos \lambda L + c_2 \sin \lambda L = 0$$

which can be written in matrix form as

$$\begin{pmatrix} 0 & \lambda \\ \cos \lambda L & \sin \lambda L \end{pmatrix} \begin{pmatrix} c_1 \\ c_2 \end{pmatrix} = \begin{pmatrix} 0 \\ 0 \end{pmatrix}.$$

This linear system has a nontrivial solution if and only if the coefficient matrix is singular. This will be the case if its determinant $f(\lambda)$ is zero. Hence we need

$$f(\lambda) \equiv -\lambda \cos \lambda L = 0$$

or $\lambda L = \frac{\pi}{2} + n\pi$ for any integer n. It follows that there are countably many values

$$\lambda_n = \frac{\frac{\pi}{2} + n\pi}{L}, \qquad n = 0, \pm 1, \pm 2, \ldots .$$

A corresponding nontrivial solution is $(c_1, c_2) = (1, 0)$. Since eigenvectors are determined only up to a multiplicative constant, we may set

$$\lambda_n = \frac{\frac{\pi}{2} + n\pi}{L}, \qquad \mu_n = -\lambda_n^2, \qquad \phi_n(x) = \cos \lambda_n x, \qquad n = 0, 1, 2, \ldots .$$

The negative integers may be ignored because they yield the same eigenvalues and eigenvectors.

Next we shall consider periodic boundary conditions.

(3.2x) $\phi(0) = \phi(L)$, $\phi'(0) = \phi'(L)$.

By inspection we find that $\lambda = \mu = 0$ is an eigenvalue with eigenfunction

$$\psi_0(x) = 1.$$

For nonzero λ the boundary conditions applied to (3.5) lead to

$$\begin{pmatrix} 1 - \cos \lambda L & \sin \lambda L \\ \lambda \sin \lambda L & \lambda(1 - \cos \lambda L) \end{pmatrix} \begin{pmatrix} c_1 \\ c_2 \end{pmatrix} = \begin{pmatrix} 0 \\ 0 \end{pmatrix}.$$

The determinant is

$$f(\lambda) \equiv \lambda(1 - \cos \lambda L)^2 + \lambda \sin^2 \lambda L = \lambda(2 - 2\cos \lambda L).$$

The determinant will be zero if and only if

$$\lambda L = 2n\pi.$$

Note that for each such λ the matrix becomes the zero matrix so that we have two linearly independent solutions for (c_1, c_2), which we may take to be $(1, 0)$ and $(0, 1)$. Hence for each nonzero eigenvalue $\mu_n = -\lambda_n^2$ there are two eigenfunctions

$$\phi_n(x) = \sin \lambda_n x, \quad \text{and}$$

$$\psi_n(x) = \cos \lambda_n x,$$

where $\lambda_n = (2n\pi)/L$. Note that formally $\psi_0(x)$ is the eigenfunction already known to us.

Finally, let us consider the general case of (3.3). If $\lambda = 0$, then substitution of (3.4) into the boundary conditions leads to the system

$$\begin{pmatrix} \alpha_2 & -\alpha_1 \\ \beta_2 & \beta_2 L + \beta_1 \end{pmatrix} \begin{pmatrix} c_1 \\ c_2 \end{pmatrix} = \begin{pmatrix} 0 \\ 0 \end{pmatrix}.$$

Under the hypotheses on the data the determinant of the coefficient matrix

$$f(0) = \alpha_2(\beta_2 L + \beta_1) + \beta_2\alpha_1$$

can vanish only if $\alpha_2 = \beta_2 = 0$ so that $\phi'(0) = \phi'(L) = 0$. Then $\phi_0(x) = 1$. For all other cases $\mu = \lambda = 0$ is not an eigenvalue. If $\lambda \neq 0$, then (3.5) must be substituted into (3.3). This leads to the following matrix equation for the coefficients c_1 and c_2:

$$\begin{pmatrix} \alpha_2 & -\alpha_1\lambda \\ \beta_2 \cos \lambda L - \lambda\beta_1 \sin \lambda L & \beta_2 \sin \lambda L + \lambda\beta_1 \cos \lambda L \end{pmatrix} \begin{pmatrix} c_1 \\ c_2 \end{pmatrix} = \begin{pmatrix} 0 \\ 0 \end{pmatrix}.$$

For a nontrivial solution the determinant $f(\lambda)$ of the coefficient matrix must be zero. This leads to the condition

$$f(\lambda) \equiv (\alpha_2\beta_2 - \lambda^2\alpha_1\beta_1)\sin \lambda L + \lambda(\alpha_2\beta_1 + \alpha_1\beta_2)\cos \lambda L = 0. \tag{3.6}$$

We note that $f(\lambda) = 0$ implies that $f(-\lambda) = 0$. If the matrix is singular, then

$$(c_1, c_2) = (\lambda\alpha_1, \alpha_2)$$

determines the coefficients of (3.5).

Simple solutions of $f(\lambda) = 0$ arise in the special cases (3.2i–iv).

$$\alpha_1 = \beta_1 = 0 \quad \text{or} \quad \alpha_2 = \beta_2 = 0$$

implies that $f(\lambda) = 0$ whenever $\sin \lambda L = 0$ or $\lambda_n = n\pi/L$, $n = 1, 2, \ldots$

$$\alpha_2 = \beta_1 = 0 \quad \text{or} \quad \alpha_1 = \beta_2 = 0$$

implies that $f(\lambda) = 0$ whenever $\cos \lambda L = 0$ or $\lambda_n = \left(\frac{\pi}{2} + n\pi\right)/L$, $n = 0, 1, \ldots$. If $\alpha_2\alpha_1 > 0$ or $\beta_2\beta_1 > 0$ (boundary conditions associated with convective heat loss, for example), then the roots of $f(\lambda)$ no longer are given in closed form but must be determined numerically. All we can say is that if $\alpha_1\beta_1 > 0$, then the roots of $f(\lambda) = 0$ approach the roots of $\sin \lambda L$ as $\lambda \to \infty$; if $\alpha_1\beta_1 = 0$, then the roots of $f(\lambda) = 0$ approach the roots of $\cos \lambda L$ as $\lambda \to \infty$. In all cases we obtain countably many values λ_n.

We can summarize the results of our discussion of the eigenvalue problem (3.1), (3.2) in the following table, to which we shall refer repeatedly as we solve partial differential equations.

Table 3.1:

Eigenvalues and eigenvectors for $\phi'' = \mu\phi$ and various boundary conditions

	Boundary Condition	$\lambda_n (\mu_n = -\lambda_n^2)$	Eigenfunction(s)	
i)	$\phi(0) = \phi(L) = 0$	$n\pi/L$	$\sin \lambda_n x,$	$n = 1, 2, \ldots$
ii)	$\phi(0) = \phi'(L) = 0$	$\left(\frac{\pi}{2} + n\pi\right)/L$	$\sin \lambda_n x,$	$n = 0, 1, \ldots$
iii)	$\phi'(0) = \phi(L) = 0$	$\left(\frac{\pi}{2} + n\pi\right)/L$	$\cos \lambda_n x,$	$n = 0, 1, \ldots$
iv)	$\phi'(0) = \phi'(L) = 0$	$n\pi/L$	$\cos \lambda_n x,$	$n = 0, 1, \ldots$
x)	$\phi(0) = \phi(L),$	$2n\pi/L$	$\sin \lambda_n x,$	$n = 1, 2, \ldots$
	$\phi'(0) = \phi'(L)$	$2n\pi/L$	$\cos \lambda_n x,$	$n = 0, 1, \ldots$

v–ix)   The remaining cases require a solution of (3.6) for the various combinations of α_i and β_j.
The corresponding eigenfunction is always

$$\phi(x) = \alpha_1 \lambda \cos \lambda x + \alpha_2 \sin \lambda x.$$

Replacing λ by $-\lambda$ does not change the eigenvalue and only changes the algebraic sign of the eigenfunction. Hence we can restrict ourselves to the positive roots of $f(\lambda) = 0$.

For illustration we show in Fig. 3.1a a plot of $f(\lambda)$ vs. λ and in Fig. 3.1b a plot of the eigenfunctions corresponding to the first two positive roots of $f(\lambda) = 0$ when $\alpha_1 = \alpha_2 = \beta_1 = \beta_2 = L = 1$. Note that these functions do not have common zeros or a common period. However, they are orthogonal in the mean square sense as guaranteed by Theorem 3.2.

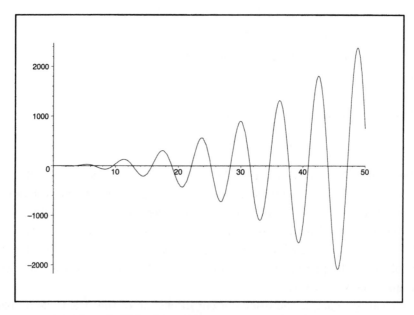

Figure 3.1: (a) Plot of $f(\lambda)$ of (3.6) for $\alpha_1 = \alpha_2 = \beta_1 = \beta_2 = L = 1$. The first two roots are $\lambda_1 = 1.30654$, $\lambda_2 = 3.67319$.

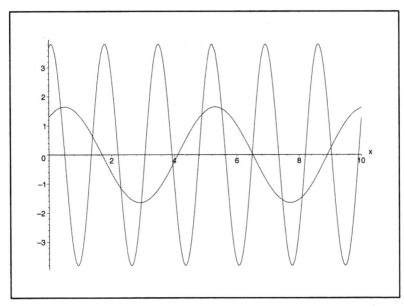

Figure 3.1: (b) Plot of $\phi_i(x) = \lambda_i \cos \lambda_i x + \sin \lambda_i x$, $i = 1, 2$.

We point out that if the sign conditions of (3.3) are relaxed, then the eigenvalues of

$$\phi'' = \mu\phi$$

remain real and the eigenfunctions corresponding to distinct eigenvalues are still orthogonal. However, some eigenvalues can be positive and eigenfunctions may now involve real exponential as well as trigonometric functions (see Exercise 3.13).

3.2   Sturm–Liouville problems for $\mathcal{L}\phi = \mu\phi$

For the simple operator $\mathcal{L}\phi \equiv \phi''$ we could exhibit countably many eigenvalues and the corresponding eigenfunctions. For more general variable coefficient operators explicit solutions are no longer available; however, for the class of Sturm–Liouville eigenvalue problems a fairly extensive theory now exists with precise statements about the existence of eigenvalues and eigenfunctions and their properties. We shall cite the aspects which are important for later applications to partial differential equations.

A typical Sturm–Liouville eigenvalue problem is given by

$$\mathcal{L}\phi \equiv (p(x)\phi'(x))' - q(x)\phi = \mu w(x)\phi \tag{3.7}$$

where p', q, and w are real continuous functions on the interval $[0, L]$, where $p(x)$ is nonnegative on $[0, L]$ and where $w(x)$ is positive except possibly at finitely many points. w is a weight function of the type introduced in Chapter 2. We shall impose on (3.7) the boundary conditions

$$p(0)[\alpha_1\phi(0) - \alpha_2\phi'(0)] = 0$$

$$p(L)[\beta_1\phi(L) + \beta_2\phi'(L)] = 0$$

where the coefficients satisfy the same conditions as in (3.3). If $p(0)$ and $p(L)$ are positive, we have exactly (3.3); but if, for example, $p(0) = 0$, then $\phi(0)$ and $\phi'(0)$ are merely assumed to exist. μ is again called the eigenvalue of problem (3.7) and $\phi \not\equiv 0$ is the corresponding eigenfunction. Assuming their existence we can readily characterize their properties.

If μ is an eigenvalue with corresponding eigenvector ϕ, then it follows from integration by parts applied to

$$\int_0^L \bar{\phi}(x)\mathcal{L}\phi(x)dx = \mu \int_0^L w(x)|\phi|^2 dx$$

that

$$p(x)\phi'(x)\overline{\phi(x)} \int_0^L - \int_0^L (p(x)|\phi'|^2 + q(x)|\phi|^2)dx = \mu \int_0^L w(x)|\phi|^2 dx.$$

The boundary conditions again guarantee that the first term on the left is real and nonpositive so that μ is real and

$$\mu \leq -\frac{\int_0^L (p(x)|\phi'|^2 + q(x)|\phi|^2)dx}{\int_0^L w(x)|\phi|^2 dx}.$$

A real eigenvalue implies that the real and imaginary parts of any complex valued eigenfunctions themselves are eigenfunctions so that again we may restrict ourselves to real vector spaces.

If we now assume that $\{\mu_m, \phi_m\}$ and $\{\mu_n, \phi_n\}$ are eigenvalue–eigenvector pairs for distinct eigenvalues, then

$$\int_0^L [\phi_m\mathcal{L}\phi_n - \phi_n\mathcal{L}\phi_m]\, dx = (\mu_n - \mu_m)\int_0^L w(x)\phi_m\phi_n\, dx.$$

Integration by parts shows that

$$\int_0^L [\phi_m\mathcal{L}\phi_n - \phi_n\mathcal{L}\phi_m]\, dx = p(x)\left[\phi_m(x)\phi_n'(x) - \phi_m'(x)\phi_n(x)\right]\Big|_0^L.$$

The boundary conditions imply that the integral vanishes. Thus we can conclude that

$$\langle \phi_m, \phi_n \rangle = \int_0^L \phi_m \phi_n w(x) dx = 0,$$

where $\langle\,,\,\rangle$ is the inner product on $L_2(0, L, w)$.

Until now we have either calculated explicitly the eigenvalues and eigenfunctions, or we have assumed their existence. The significance of the Sturm–Liouville problem (3.7) is that such assumption is justified by the following theorem [16].

Theorem 3.3 *Assume that in* (3.7) *the coefficients are continuous and that* $p, w > 0$ *on* $[0, L]$. *Then there are countably many real decreasing eigenvalues* $\{\mu_n\}$ *and eigenfunctions* $\{\phi_n\}$ *with*

$$\lim_{n \to \infty} \mu_n \to -\infty.$$

For each eigenvalue there are at most two linearly independent eigenfunctions which may be chosen to be orthogonal. All eigenfunctions constitute an orthogonal basis of the inner product space $L_2(0, L, w)$ *and for any* $f \in L_2(0, L, w)$

$$\lim_{N \to \infty} \|f - P_N f\| = 0$$

where $P_N f = \sum_{n=1}^N \frac{\langle f, \phi_n \rangle}{\langle \phi_n, \phi_n \rangle} \phi_n$ *is the orthogonal projection of* f *onto* span$\{\phi_1, \ldots, \phi_N\}$.

The theorem asserts that $L_2(0, L, w)$ contains countably many orthogonal elements $\{\phi_n\}$. Since any finite number of these elements are necessarily linearly independent, we see that $L_2(0, L, w)$ is an infinite-dimensional inner product space. The definition of basis in Chapter 2 must be broadened to apply in an infinite-dimensional vector space X. Let $\{x_n\}$ be a sequence of elements of a normed linear space X. We say that $\{x_n\}$ is a basis of X if any finite set of these vectors is linearly independent and for any $x \in X$, there is a sequence of scalars $\{\alpha_n\}$ such that

$$\lim_{N \to \infty} \left\| x - \sum_{n=1}^N \alpha_n x_n \right\| = 0.$$

For the orthogonal basis of Theorem 3.3 the linear combinations may be taken to be the orthogonal projections.

Throughout the following chapters we are going to approximate given functions by their orthogonal projections onto the span of finitely many eigenfunctions of Sturm–Liouville problems. It is reassuring to know that the approximations converge as we take more and more eigenfunctions. Unfortunately, such convergence can be painfully slow. Let us take the function $f(x) \equiv 1$ in $L_2(0,1)$ and project it into the span of eigenfunctions $\{\sin n\pi x\}$ of case i) in Table 3.1. From Theorem 3.3 we know that $\|1 - P_N 1\| \to 0$ as $N \to \infty$. A simple calculation yields

$$P_N 1 = 2 \sum_{n=1}^{N} \frac{1 - \cos(n\pi)}{n\pi} \sin n\pi x.$$

Computed values (in single precision) are

N	$\|1 - P_N 1\|_{L_2(0,1)}$
10	.20099
100	.06366
1000	.02013
10000	.00815

That is about as close as we can come numerically. On the other hand, for $f(x) \equiv x(1-x)$ we obtain

N	$\|x(1-x) - P_N x(1-x)\|_{L_2(0,1)}$
10	.00019
20	.00009

The next chapter gives some insight into why $f(x) \equiv 1$ is difficult and $f(x) \equiv x(1-x)$ is easy to approximate in this setting. Here we merely would like to point out that when in later examples we solve the Dirichlet problem

$$\Delta u = 1, \qquad (x,y) \in D$$
$$u = 0, \qquad (x,y) \in \partial D,$$

then the simple source term $F(x,y) \equiv 1$ generally will make the mechanics of solving the problem easier but does not favor convergence of the approximate solution to the analytic solution.

If $p \geq 0$ and $p(0)p(1) = 0$ or $w = 0$ at finitely many points, then we have a singular Sturm–Liouville problem. The eigenvalue problem for Bessel's equation in Chapter 6 is a typical example of a singular problem. The theory for such problems becomes more complicated but in the context of separation of variables it is safe to assume that p and w are such that the conclusion of Theorem 3.3 remains valid. In particular, this means that we always expect that

$$\lim_{N \to \infty} \|F - P_N F\| \equiv \lim_{N \to \infty} \left\| F - \sum_{n=1}^{N} \gamma_n \phi_n \right\| = 0$$

where F is an arbitrary function in $L_2(0, L, w)$ and γ_n is the Fourier coefficient

$$\gamma_n = \frac{\langle F, \phi_n \rangle}{\langle \phi_n, \phi_n \rangle}$$

with

$$\langle f, g \rangle = \int_0^L f(x)g(x)w(x)dx.$$

Next we observe that the general eigenvalue problem

$$\mathcal{L}\phi \equiv a(x)\phi'' + b(x)\phi' + c(x)\phi = \mu\phi \tag{3.8}$$

can be put into Sturm–Liouville form (3.7) if there is a weight function $w(x)$ such that

$$[a(x)\phi'' + b(x)\phi' + c(x)\phi]w(x) = \mu\phi w(x)$$

can be written in the form of (3.7). A comparison shows that we would need

$$a(x)w(x) = p(x)$$

$$b(x)w(x) = p'(x)$$

so that

$$(aw)' = bw.$$

If $a(x) \neq 0$ on $[0, L]$, then we can write

$$(aw)' = \frac{b(x)}{a(x)}(aw)$$

and find that for any x_0

$$w(x) = \frac{1}{a(x)} e^{\int_{x_0}^x \frac{b(s)}{a(s)}\,ds}$$

is an admissible weight function. Hence eigenfunctions of (3.8) can be found such that they form an orthogonal basis of $L_2(0, L, w)$. For example, consider the eigenvalue problem

$$\phi'' + \frac{2}{x}\phi' = \mu\phi \tag{3.9}$$

$$\phi'(0) = \phi'(1) = 0.$$

We see that

$$w(x) = e^{\int_1^x \frac{2}{s}\,ds} = x^2$$

and that the problem can be rewritten as

$$(x^2\phi')' = \mu x^2 \phi$$

$$\phi'(0) = \phi'(1) = 0.$$

The above discussion immediately yields that eigenvalues (should they exist) are negative and that eigenfunctions corresponding to distinct eigenvalues are orthogonal in $L(0, 1, x^2)$.

Throughout this text eigenvalue problems like (3.8) are useful for solving partial differential equations only if one can actually compute the eigenvalues and eigenfunctions. As a rule that is a difficult if not impossible task since there is no general recipe for solving ordinary differential equations with variable coefficients. When confronted with an unfamiliar differential equation, about the only choice is to check whether the equation is listed in handbooks of solutions for ordinary differential equations such as [12], [18]. Fortunately, (3.9) appears as equation 2.101 in [12] and is solvable by elementary means. We rewrite the equation as

$$(x\phi)'' = -\lambda^2(x\phi)$$

so that

$$x\phi(x) = c_1 \cos \lambda x + c_2 \sin \lambda x.$$

It is straightforward to verify that

$$\lim_{x \to 0} \left(\frac{\cos \lambda x}{x} \right)' \quad \text{does not exist so that } c_1 = 0,$$

while

$$\lim_{x \to 0} \left(\frac{\sin \lambda x}{x} \right)' = 0.$$

Hence

$$\phi_n(x) = \frac{\sin \lambda_n x}{x}$$

where $\{\lambda_n\}$ are the solutions $\phi'_n(1) = 0$ which leads to

$$f(\lambda) \equiv \lambda \cos \lambda - \sin \lambda = 0.$$

Since $f(n\pi)f((n + 1)\pi) < 0$, there is a root in each subinterval $(n\pi, (n + 1)\pi)$, but its value can only be found numerically.

Many applications of separation of variables lead to ordinary differential equations which are solved in terms of Bessel functions (see Examples 6.10, 7.5, 8.5, 9.8). Bessel functions belong to the class of "special functions" studied in such texts as [1] (for real arguments) and [13] (for complex arguments). Because of the ubiquity of Bessel functions, we include here for easy reference a general second order ordinary differential equation which can be solved in terms of Bessel functions. We cite from [21].

If $(1-a)^2 \geq 4c$ and if neither d, p, nor q is zero, then, except in the obvious special cases when it reduces to Euler's equation, the differential equation

$$x^2 u'' + x(a + 2bx^p)u' + [c + dx^{2q} + b(a + p - 1)x^p + b^2 x^{2p}]u = 0 \qquad (3.10)$$

has the general solution

$$u(x) = x^\alpha e^{-\beta x^p}[c_1 J_\nu(\lambda x^q) + c_2 Y_\nu(\lambda x^q)]$$

where

$$\alpha = \frac{1-a}{2} \qquad \beta = \frac{b}{p} \qquad \lambda = \frac{\sqrt{|d|}}{q} \qquad \nu = \frac{\sqrt{(1-a)^2 - 4c}}{2q}.$$

If $d < 0$, J_ν and Y_ν are to be replaced by I_ν and K_ν, respectively. If ν is not an integer, Y_ν and K_ν can be replaced by $J_{-\nu}$ and $I_{-\nu}$ if desired.

The definitions and properties of the various functions just cited are discussed in the above sources (see also Example 6.10). The manipulation and evaluation of these functions have become routine in programming environments like Maple, Mathematica, and Matlab.

3.3   A Sturm–Liouville problem with an interface

We shall conclude our discussion of eigenvalues for equation (3.8) by considering the following generalization of the simple equation (3.1):

$$a(x)\phi'' = \mu\phi \quad \text{on } (0, X) \cup (X, L)$$

where

$$a(x) = \begin{cases} a_1, & 0 < x < X \\ a_2, & X < x < L \end{cases}$$

for given positive constants a_i and a given interface X. At $x = 0$ and $x = L$ the function $\phi(x)$ may be subject to any of the boundary conditions of Table 3.1, but for definiteness we shall choose here

$$\phi(0) = \phi'(L) = 0.$$

At $x = X$ the eigenfunction is required to satisfy an interface condition of the form

$$\phi(X^-) = \phi(X^+)$$
$$A_1 \phi'(X^-) = A_2 \phi'(X^+)$$

where A_1 and A_2 are given positive constants. Interface conditions like these arise when the method of separation of variables is applied in composite media (see Example 6.11).

Let us introduce the piecewise constant weight function

$$w(x) = \begin{cases} a_2 A_1, & 0 < x < X \\ a_1 A_2, & X < x < L \end{cases}$$

and the inner product

$$\langle f, g \rangle = \int_0^L f(x) g(x) w(x) dx.$$

Then

$$\int_0^L \bar{\phi}(x) a(x) \phi''(x) w(x) dx = \mu \int_0^L |\phi(x)|^2 w(x) dx.$$

We integrate the left side by parts over $[0, X]$ and $[X, L]$ and obtain

$$\int_0^X \bar{\phi} a(x) \phi'' w(x) dx = \int_0^X -a_1 |\phi'|^2 w(x) dx + a_1 \bar{\phi}(x) \phi'(x) a_2 A_1 \Big|_0^X$$

$$\int_X^L \bar{\phi} a(x) \phi'' w(x) dx = \int_X^L -a_2 |\phi'|^2 w(x) dx + a_2 \bar{\phi}(x) \phi'(x) a_1 A_2 \Big|_X^L.$$

The boundary and interface conditions imply that the boundary and interface terms drop out or cancel. Hence

$$\mu \int_0^L |\phi(x)|^2 w(x) dx = - \int_0^L a(x) |\phi'(x)|^2 w(x) dx$$

so that an eigenvalue of this problem is real and nonpositive and has a real eigenfunction. Similarly, it follows from integration by parts applied to

$$\int_0^L a(x) [\phi_m'' \phi_n - \phi_n'' \phi_m] w(x) dx = (\mu_m - \mu_n) \int_0^L \phi_m \phi_n w(x) dx$$

that the left integral vanishes so that eigenfunctions corresponding to distinct eigenvalues are orthogonal in $L_2(0, L, w)$. Eigenvalues and eigenfunctions can be found explicitly. We observe first that

$$\mu = 0$$

would lead to

$$\phi(x) = \begin{cases} c_1 + c_2 x, & 0 < x < X \\ d_1 + d_2 x, & X < x < L. \end{cases}$$

The boundary conditions require that $c_1 = d_2 = 0$. The interface conditions

$$c_2 X = d$$
$$A_1 c_2 = 0$$

imply that $c_2 = d_1 = 0$ and show that there is no nonzero solution corresponding to $\mu = 0$. Let us then write

$$\mu = -\lambda^2 \quad \text{for } \lambda \neq 0.$$

Then any solution of our interface problem must be of the form

$$\phi(x) = \begin{cases} c_1 \cos \frac{\lambda}{\sqrt{a_1}} x + c_2 \sin \frac{\lambda}{\sqrt{a_1}} x, & 0 < x < X \\ d_1 \cos \frac{\lambda}{\sqrt{a_2}} x + d_2 \sin \frac{\lambda}{\sqrt{a_2}} x, & X < x < L. \end{cases}$$

The boundary conditions allow us to write

$$\phi(x) = \begin{cases} c \sin \frac{\lambda}{\sqrt{a_1}} x \\ d \cos \frac{\lambda}{\sqrt{a_2}} (L - x). \end{cases}$$

The interface conditions lead to

$$\begin{pmatrix} \sin \frac{\lambda}{\sqrt{a_1}} X & -\cos \frac{\lambda}{\sqrt{a_2}} (L - X) \\ A_1 \frac{\lambda}{\sqrt{a_1}} \cos \frac{\lambda}{\sqrt{a_1}} X & -A_2 \frac{\lambda}{\sqrt{a_2}} \sin \frac{\lambda}{\sqrt{a_2}} (L - X) \end{pmatrix} \begin{pmatrix} c \\ d \end{pmatrix} = \begin{pmatrix} 0 \\ 0 \end{pmatrix}.$$

A nonzero solution results if the determinant $f(\lambda)$ of the coefficient matrix is zero. Hence we need roots of

$$f(\lambda) \equiv -A_2 \frac{\lambda}{\sqrt{a_2}} \sin \frac{\lambda}{\sqrt{a_1}} X \sin \frac{\lambda}{\sqrt{a_2}} (L - x)$$
$$+ A_1 \frac{\lambda}{\sqrt{a_1}} \cos \frac{\lambda}{\sqrt{a_1}} X \cos \frac{\lambda}{\sqrt{a_2}} (L - X) = 0.$$

This expression can be rewritten in the form

$$f(\lambda) = +A_1 \frac{\lambda}{\sqrt{a_1}} \cos \left[\frac{\lambda}{\sqrt{a_1}} X + \frac{\lambda}{\sqrt{a_2}} (L - X) \right]$$
$$+ \lambda \left(\frac{A_1}{\sqrt{a_1}} - \frac{A_2}{\sqrt{a_2}} \right) \sin \frac{\lambda}{\sqrt{a_1}} X \sin \frac{\lambda}{\sqrt{a_2}} (L - X) = 0.$$

We see that for $a_1 = a_2$ and $A_1 = A_2$ the eigenvalue condition reduces to that of ii) in Table 3.1. This is the correct behavior because the solution is now twice continuously differentiable at X so that the location of X should not have any influence. The eigenfunction corresponding to any nonzero root of this equation is

$$\phi(x) = \begin{cases} \cos \frac{\lambda}{\sqrt{a_2}} (L - X) \sin \frac{\lambda}{\sqrt{a_1}} x, & 0 < x < X \\ \sin \frac{\lambda}{\sqrt{a_1}} X \cos \frac{\lambda}{\sqrt{a_2}} (L - x), & X < x < L. \end{cases}$$

For $a_1 = a_2$ and $A_1 = A_2$ trigonometric addition formulas yield a scalar multiple of the eigenfunction of ii) in Table 3.1, as expected. We observe that $f(-\lambda) = -f(\lambda)$ and that $\phi(x)$ only changes sign if $\lambda \to -\lambda$. Hence again we only need to consider positive roots of $f(\lambda) = 0$.

We shall conclude our glimpse into the world of Sturm–Liouville problems by pointing out that the theory extends to eigenvalue problems for partial differential equations. For example, suppose we consider the eigenvalue problem for Laplace's equation

$$\Delta\phi = \phi_{xx} + \phi_{yy} = \mu\phi, \qquad (x,y) \in D$$

$$\alpha_1\phi + \alpha_2 \frac{\partial\phi}{\partial n} = 0, \qquad (x,y) \in \partial D$$

where α_1 and α_2 are piecewise smooth functions that satisfy

$$\alpha_1\alpha_2 \geq 0, \qquad \alpha_1^2 + \alpha_2^2 > 0,$$

and where D is a domain in R_2 with smooth boundary ∂D. Let $\{\mu, \phi\}$ be an eigenvalue and eigenvector; then

$$\int_D \bar\phi \Delta\phi \, d\vec{x} = \mu \int_D |\phi^2| d\vec{x}$$

with $d\vec{x} = dx\,dy$. An application of the divergence theorem yields

$$\int_D \bar\phi \Delta\phi \, d\vec{x} = \int_D \nabla\cdot\bar\phi\nabla\phi\, d\vec{x} - \int_D |\nabla\phi|^2|d\vec{x} = \int_{\partial D} \bar\phi\nabla\phi\cdot\vec{n}\, ds - \int_D |\nabla\phi|^2 d\vec{x}.$$

The boundary integral is nonpositive because

$$\frac{\alpha_1 + \alpha_2}{\alpha_1 + \alpha_2} \bar\phi\nabla\phi\cdot\vec{n} = \frac{-\left[\alpha_1|\phi|^2 + \alpha_2\left|\frac{\partial\phi}{\partial n}\right|^2\right]}{\alpha_1 + \alpha_2}.$$

Thus the eigenvalues of the Laplacian are nonpositive. Moreover, $\mu = 0$ would require that $\nabla\phi = 0$ throughout D so that ϕ is constant in D. If $\alpha_1 > 0$ at one point on ∂D, then necessarily $\phi = 0$ at that point so that $\phi \equiv 0$ throughout D. In this case $\mu = 0$ is not an eigenvalue. If $\alpha_1 \equiv 0$ everywhere on ∂D, then $\mu = 0$ is an eigenvalue with corresponding eigenfunction $\phi \equiv 1$. Finally, we apply the divergence theorem to the left side of

$$\int_D (\phi_n\Delta\phi_m - \phi_m\Delta\phi_n)d\vec{x} = (\mu_m - \mu_n)\int_D \phi_m\phi_n d\vec{x}$$

and obtain

$$\int_D (\phi_n\Delta\phi_m - \phi_m\Delta\phi_n)d\vec{x} = \int_D \nabla\cdot(\phi_n\nabla\phi_m - \phi_m\nabla\phi_n)d\vec{x}$$

$$= \int_{\partial D} \left(\frac{\alpha_1 + \alpha_2}{\alpha_1 + \alpha_2}\right)\left(\phi_n\frac{\partial\phi_m}{\partial n} - \phi_m\frac{\partial\phi_n}{\partial n}\right)ds = 0$$

because

$$\alpha_1 \phi_m + \alpha_2 \frac{\partial \phi_m}{\partial n} = \alpha_1 \phi_n + \alpha_2 \frac{\partial \phi_n}{\partial n} = 0 \quad \text{on } \partial D.$$

We can conclude that the eigenfunctions corresponding to distinct eigenvalues satisfy

$$\int_D \phi_m \phi_n d\vec{x} = 0;$$

i.e., they are orthogonal with respect to the inner product

$$\langle f, g \rangle = \int_D f(\vec{x}) g(\vec{x}) d\vec{x}.$$

Finally, we mention that Theorem 3.3 has its analogue for multidimensional Sturm–Liouville problems. The consequence for the later chapters of this text is that a function F which is square integrable over the domain D (i.e., $F \in L_2(D)$) can be approximated in terms of the eigenfunctions of the Laplacian such that

$$\left\| F - \sum_{n=1}^{N} \gamma_n \phi_n \right\| \to 0 \quad \text{as } N \to \infty$$

where

$$\gamma_n = \frac{\langle F, \phi_n \rangle}{\langle \phi_n, \phi_n \rangle}.$$

The formalism stays the same; only the meaning of the inner product changes with the setting of the problem.

Exercises

3.1) Find all eigenvalues and eigenvectors of the following problems:
   i) $\phi'' = \mu \phi$
      $\phi(0) = 0$, $\phi(1) = 0$.
   ii) $\phi'' = \mu \phi$
      $\phi'(0) = 0$, $\phi'(1) = 0$.
   iii) $\phi'' + \phi' = \mu \phi$
      $\phi(0) = 0$, $\phi(2) = 0$.

3.2) Consider

$$\phi'' = \mu \phi$$

$$\phi(1) = 0, \qquad \phi'(L) = 0.$$

Find all L such that $\mu = -3$ is an eigenvalue.

3.3) Solve the eigenvalue problem

$$\phi'' = \mu\phi$$

on each of the following subspaces $C^2(0, L)$ with $L > 0$:
  i) $M = \{\phi \in C^2[0, L], \phi(0) = 0\}$.
  ii) $M = \{\phi \in C^2[0, L], \phi'(0) = 0\}$.
  iii) $M = \{\phi \in C^2[0, L], \phi(0) = \phi(L/2) = \phi(L)\}$.
  iv) $M = \{\phi \in C^2[0, L], \phi(0) = 0, \int_0^L \phi(x)dx = 0\}$.
  v) $M = \{\phi \in C^2[0, L], \int_0^L x\phi(x)dx = 0\}$.

3.4) Verify that the boundary conditions (3.2x) imply that for any two eigen-functions $\{\phi_m, \phi_n\}$

$$\int_0^L [\phi_n \mathcal{L}\phi_m - \phi_m \mathcal{L}\phi_n]\, dx = 0.$$

3.5) Verify that the boundary conditions (3.3) imply that for any two eigenfunc-tions $\{\phi_m, \phi_n\}$

$$\int_0^L [\phi_n \mathcal{L}\phi_m - \phi_m \mathcal{L}\phi_n]\, dx = 0.$$

3.6) For the boundary condition (3.2x) verify that for $\mu = 0$ the determinant of the coefficient matrix is zero and that $\phi(x) = c \neq 0$ is the corresponding eigenfunction.

3.7) For the boundary condition (3.3) verify that $\mu = 0$ is an eigenvalue if and only if $\alpha_1 = \beta_1 = 0$.

3.8) Verify all eigenvalues and eigenfunctions of Table 3.1.

3.9) Find all μ for which there exists a nontrivial solution of

$$\phi'' + \phi = 0$$
$$\phi(0) - \phi'(0) = 0$$
$$\phi(1) + \mu\phi'(1) = 0.$$

3.10) Convert

$$\phi''(x) + x^2\phi'(x) = \mu\phi(x)$$
$$\phi(0) = 0, \qquad \phi'(1) = 0$$

into a regular Sturm–Liouville problem and find the inner product associated with it.

3.11) Let $\{-\lambda^2, \phi\}$ be an eigenvalue eigenvector pair of the problem

$$\phi'' = -\lambda^2 \phi$$

$$\phi(0) = 0, \quad \phi(L) + \phi'(L) = 0.$$

i) Use (3.6) to show that

$$\int_0^L \phi^2(x)dx = \frac{L}{2} + \frac{\cos^2 \lambda L}{2}.$$

ii) Show by integrating that $\langle \phi_m, \phi_n \rangle = 0$ for two eigenfunctions with distinct eigenvalues.

3.12) Find the eigenvalues and eigenfunctions of the interface problem

$$\phi''(x) = \mu\phi(x)$$

$$\phi(0) = 0, \qquad \phi(3) = 0$$

$$\phi(1-) = \phi(1+)$$

$$\phi'(1-) = 2\phi'(1+).$$

3.13) i) Show that the eigenvalue problem

$$\phi'' = \mu\phi$$

$$\phi'(0) = 0, \qquad \phi(1) = \phi'(1)$$

has exactly one positive eigenvalue and countably many negative eigenvalues.

ii) Compute numerically the positive eigenvalue and the corresponding eigenfunction.

iii) Compute the first negative eigenvalue and the corresponding eigenfunction.

iv) Show by direct integration that the eigenfunctions of ii) and iii) are orthogonal.

v) Why does the existence of a positive eigenvalue not contradict Theorem 3.2?

3.14) Find the eigenvalues and eigenfunctions of

$$\phi^{(iv)}(x) = \mu\phi(x)$$
$$\phi(0) = \phi(1) = 0$$
$$\phi''(0) = \phi''(1) = 0.$$

3.15) Find the general solution of

$$(x^r u')' + (ax^s + bx^{r-2})u = 0$$

when
  i) $(1-r)^2 \geq 4b$,
  ii) $r = 2$, $s = b = 0$,
  iii) $a = 0$.

Chapter 4

Fourier Series

4.1 Introduction

In the discussion of Sturm–Liouville problems we saw that the boundary value problem

$$\phi''(x) = \mu\phi(x), \qquad -L \leq x \leq L;$$
$$\phi(-L) = \phi(L), \qquad\qquad (4.1)$$
$$\phi'(-L) = \phi'(L)$$

has eigenvalues $\mu_n = -\lambda_n^2$, where $\lambda_n = n\pi/L$, for $n = 0, 1, 2, \ldots$ and corresponding eigenfunctions

$$\left\{ 1, \cos\frac{\pi}{L}x, \sin\frac{\pi}{L}x, \cos 2\frac{\pi}{L}x, \sin 2\frac{\pi}{L}x, \ldots, \cos n\frac{\pi}{L}x, \sin n\frac{\pi}{L}x, \ldots \right\}.$$

The Sturm–Liouville theory assures that for $f \in L_2(-L, L)$, the sequence of orthogonal projections $P_N f$ of the function f onto $M = \operatorname{span}\left\{ 1, \cos\frac{\pi}{L}x, \sin\frac{\pi}{L}x, \ldots, \cos N\frac{\pi}{L}x, \sin N\frac{\pi}{L}x \right\}$ converges in the mean to f. More specifically,

$$P_N f = a_0 + \sum_{n=1}^{N} \left(a_n \cos n\frac{\pi x}{L} + b_n \sin n\frac{\pi x}{L} \right),$$

where

$$a_n = \frac{\int\limits_{-L}^{L} f(t) \cos\frac{n\pi t}{L}\, dt}{\int_{-L}^{L} \cos^2\frac{n\pi t}{L}\, dt} = \begin{cases} \frac{1}{2L}\int_{-L}^{L} f(t)\, dt, & n = 0 \\ \frac{1}{L}\int_{-L}^{L} f(t)\cos\frac{n\pi t}{L}\, dt, & n \geq 1 \end{cases}$$

$$b_n = \frac{\int\limits_{-L}^{L} f(t) \sin\frac{n\pi t}{L}\, dt}{\int_{-L}^{L} \sin^2\frac{n\pi t}{L}\, dt} = \frac{1}{L}\int_{-L}^{L} f(t)\sin\frac{n\pi t}{L}\, dt, \qquad n \geq 1,$$

67

is the orthogonal projection of f onto M_N and the series

$$a_0 + \sum_{n=1}^{\infty} \left(a_n \cos n\frac{\pi x}{L} + b_n \sin n\frac{\pi x}{L} \right)$$

converges in the mean to f. For this series, however, a great deal more is known about its approximating properties than just mean square convergence. The series is called the Fourier series of f, named for the French mathematician and physicist Joseph Fourier (1768–1830), who introduced it in his solutions of the heat equation. The fact that a series is the Fourier series of a function f is usually indicated by the notation

$$f \sim a_0 + \sum_{n=1}^{\infty} \left(a_n \cos n\frac{\pi x}{L} + b_n \sin n\frac{\pi x}{L} \right).$$

The coefficients a_n and b_n are called the Fourier coefficients of f.

The theory of such series is the basis of trigonometric approximations to continuous and discontinuous periodic functions widely used in signal recognition and data compression as well as in diffusion and vibration studies. This chapter introduces the mathematics of Fourier series needed later on for eigenfunction expansions.

Many of the Sturm–Liouville problems discussed, including (4.1), lead to periodic eigenfunctions — mostly various linear combinations of sines and cosines. The orthogonal projection $P_N f$ of a function f defined on an interval onto the span M_N of such functions is thus defined on all of R. Moreover, if the eigenfunctions which generate M_N have a common period, then $P_N f$ is periodic. Most of the results to follow are stated for periodic functions f rather than for functions defined on an interval $[-L, L]$. At first sight this appears to be a serious restriction, but upon reflection we see this is not so and actually provides more generality to the theory. If we are concerned, as we usually are, with a function f defined only on an interval, we apply our results for periodic functions to the so-called periodic extension of f. This is simply a function $\tilde{f}$ defined on all of R which has period $2L$, which agrees with f on the interval $(-L, L)$, and which takes on the value $\tilde{f}(-L) = \tilde{f}(L) = \frac{f(L) + f(-L)}{2}$.

Example 4.1 Let f be the function defined on $[-\pi, \pi]$ by $f(x) = x$. Here is a picture of the periodic extension of f. In this case $\tilde{f}(\pi) = 0$. See Fig. 4.1.

4.2   Convergence

Suppose X is an inner product space, and $\{\varphi_1, \varphi_2, \varphi_3, \ldots\}$ is a linearly independent sequence of elements of X and $f \in X$. We have seen in Chapter 2 that if

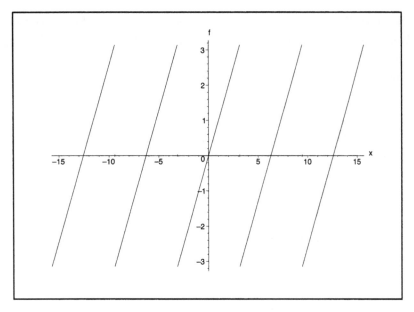

Figure 4.1: Periodic extension of $f(x) = x$, $x \in [-\pi, \pi]$.

$M_n = \text{span}\{\varphi_1, \varphi_2, \ldots, \varphi_n\}$, then the orthogonal projection $P_n f$ of f onto M_n is the best approximation of f in the space M_n. Since $M_1 \subset M_2 \subset \cdots \subset M_n \subset \cdots$, it is clear that

$$\|f - P_1 f\| \geq \|f - P_2 f\| \geq \cdots \geq \|f - P_n f\| \geq \cdots \geq 0,$$

and a natural question is, under what circumstances does the sequence $(\|f - P_n f\|)$ converge to 0? Let us see.

Definition In a normed space, a sequence $\{f_n\}$ is said to converge in the mean (or in norm) to f if $\lim_{n\to\infty} \|f - f_n\| = 0$.

Definition Suppose $B = \{\varphi_1, \varphi_2, \varphi_3, \ldots\}$ is a linearly independent set of elements of an inner product space X. If for every $f \in X$ it is true that for any $\varepsilon > 0$ there is an N so that $\|f - P_N f\| < \varepsilon$, where $P_N f$ is the orthogonal projection of f onto $M_N = \text{span}\{\varphi_1, \varphi_2, \ldots, \varphi_N\}$, then B is a basis for X.

Example 4.2 Consider the space $C[-1, 1]$ of all continuous functions on the interval $[-1, 1]$ with the norm induced by the "usual" inner product

$$\langle f, g \rangle = \int_{-1}^{1} f(x)g(x)dx.$$

The classical Weierstrass approximation theorem tells us that for any continuous f, given $\varepsilon > 0$, there is a polynomial p so that $|f(x) - p(x)| < \varepsilon$. This implies $\|f - p\| = \sqrt{2}\,\epsilon$ and thus $B = \{1, x, x^2, \ldots\}$ is a basis.

For computational efficiency we want to have

$$P_n f = \alpha_1 \varphi_1 + \alpha_2 \varphi_2 + \cdots + \alpha_n \varphi_n,$$

where the coefficients α_j do not depend on n. In this way, the sequence of projections $(P_n f)$ can be written nicely as a series

$$\sum_{i=1}^{\infty} \alpha_i \varphi_i.$$

This can be ensured by requiring that $\{\varphi_1, \varphi_2, \varphi_3, \ldots\}$ be orthogonal; that is, $\langle \varphi_j, \varphi_k \rangle = 0$ for $j \neq k$. Then, as we have seen in Chapter 2

$$\alpha_i = \frac{\langle f, \varphi_i \rangle}{\langle \varphi_i, \varphi_i \rangle}.$$

Given an orthogonal set $\{\varphi_1, \varphi_2, \varphi_3, \ldots\}$, if for each n we replace φ_i by $\widehat{\varphi}_i = \frac{1}{\|\varphi_i\|}\varphi_i$, we obtain an orthogonal set $\{\widehat{\varphi}_1, \widehat{\varphi}_2, \widehat{\varphi}_3, \ldots\}$ each element of which has norm 1 and which is equivalent to the original one in the sense that span $\{\varphi_1, \varphi_2, \ldots, \varphi_n\} = $ span $\{\widehat{\varphi}_1, \widehat{\varphi}_2, \widehat{\varphi}_3, \ldots\}$. Such a set is said to be orthonormal. If $\{\varphi_1, \varphi_2, \varphi_3, \ldots\}$ is an orthonormal set, then

$$P_n f = \sum_{i=1}^{n} \alpha_i \varphi_i,$$

where

$$\alpha_i = \langle f, \varphi_i \rangle.$$

Proposition 4.3 *If $\{\varphi_1, \varphi_2, \varphi_3, \ldots\}$ is an orthonormal basis and $P_n f = \sum_{i=1}^{n} \alpha_i \varphi_i$, then the series $\sum_{i=1}^{\infty} |\alpha_i|^2$ converges and*

$$\sum_{i=1}^{\infty} |\alpha_i|^2 \leq \|f\|^2.$$

Proof. For all n, we have

$$0 \leq \|f - P_n f\|^2 = \langle f - P_n f, f - P_n f \rangle = \langle f - P_n f, f \rangle$$

$$= \|f\|^2 - \sum_{i=1}^{n} \alpha_i \langle \varphi_i, f \rangle = \|f\|^2 - \sum_{i=1}^{n} \alpha_i \overline{\alpha}_i.$$

Thus

$$\sum_{i=1}^{n} |\alpha_i|^2 \leq \|f\|^2,$$

from which the proposition follows at once.

Corollary 4.4 $\lim_{n\to\infty} \alpha_n = \lim_{n\to\infty} \langle f, \varphi_n \rangle = 0.$

Corollary 4.5 $f = \sum_{i=1}^{\infty} \alpha_i \varphi_i$ *if and only if* $\|f\|^2 = \sum_{i=1}^{\infty} |\alpha_i|^2.$

The inequality $\sum_{i=1}^{\infty} |\alpha_2|^2 \leq \|f\|^2$ is known as Bessel's inequality, while the equation $\sum_{i=1}^{\infty} |\alpha_i|^2 = \|f\|$ is called Parseval's identity.

In virtually all our applications, the inner product spaces in which we are interested will be spaces of real- or complex-valued functions. There are thus other types of convergence to be considered. A review of these is in order. In addition to convergence in the mean, we have pointwise convergence and uniform convergence.

Definition A sequence $\{f_n\}$ of functions all with domain D is said to converge pointwise to the function f if for each $x \in D$, it is true that $f_n(x) \to f(x)$.

Definition A sequence $\{f_n\}$ of functions all with domain D is said to converge uniformly to the function f if given an $\varepsilon > 0$, there is an integer N so that $|f_n(x) - f(x)| < \varepsilon$ for all $n \geq N$ and all $x \in D$.

A few examples will help illuminate these definitions.

Example 4.6 a) For each positive integer $n \geq 2$ and for $x \in [0, 1]$, let

$$f_n(x) = \begin{cases} n^2 x & \text{if } 0 \leq x \leq 1/n \\ -n^2(x - \frac{1}{n}) + n & \text{if } 1/n < x \leq 2/n \\ 0 & \text{if } 2/n < x \leq 1. \end{cases}$$

Thus, what f_n looks like is shown in Fig. 4.2.

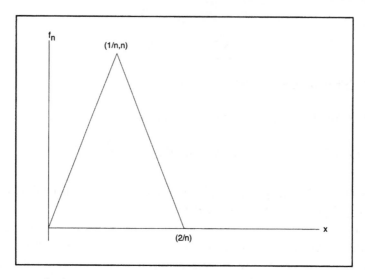

Figure 4.2: $\{f_n\}$ converges pointwise, but not uniformly or in the mean.

It should be clear that for each $x \in [0, 1]$, we have $\lim_{n \to \infty} f_n(x) = 0$. Thus the sequence $\{f_n\}$ converges *pointwise* to the function $f(x) = 0$. Note that this sequence does not, however, converge *uniformly* to $f(x) = 0$. In fact, for every n, there is an $x \in [0, 1]$ such that $|f_n(x) - f(x)| \geq n$.

Note that

$$\int_0^1 (f_n(x))^2 \, dx = \frac{2}{3}n,$$

which tells us that $\{f_n\}$ does not converge in the mean to f (with weight function $w(x) \equiv 1$, of course).

b) Next, for each integer $n \geq 1$ and $x \in [0, 1]$, let

$$f_n(x) = \begin{cases} n - n^4 x & \text{if } 0 \leq x \leq 1/n^3 \\ 0 & \text{if } 1/n^3 < x \leq 1. \end{cases}$$

A picture is given in Fig. 4.3.

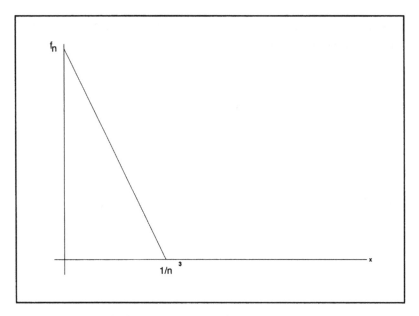

Figure 4.3: $\{f_n\}$ converges in the mean, but not pointwise.

It is clear that $\int_0^1 (f_n(x))^2 dx = \frac{1}{3n}$, and so our sequence converges in the mean to $f(x) = 0$. (Again, we have $w(x) \equiv 1$.) Clearly it does not converge to f pointwise or uniformly.

We see next that uniform convergence is the nicest of the three.

Theorem 4.7 *Suppose the sequence $\{f_n\}$ of functions in $L_2(D, w)$, with domain $D = [a, b]$, converges uniformly to the function f. Then $\{f_n\}$ converges pointwise to f and also converges in the mean to f.*

Proof. It is obvious that the sequence converges to f pointwise. For convergence in the mean, let $K = \max \{w(x) : x \in D\}$. Let $\varepsilon > 0$ and choose n sufficiently large to ensure that

$$|f_n(x) - f(x)| < \sqrt{\frac{\varepsilon}{K(b-a)}} \text{ for all } x \in [a, b].$$

Then

$$\int_a^b |f_n(x) - f(x)|^2 \, w(x) dx < \int_a^b \frac{\varepsilon}{b-a} dx = \varepsilon.$$

Hence, $\{f_n\}$ converges in the mean to f.

Theorem 4.8 *Suppose $\{f_n\}$ is a sequence of continuous functions on a domain D that converges uniformly to the function f. Then f is also continuous on D.*

Proof. To show that f is continuous, let $x_0 \in D$ and let $\varepsilon > 0$ be given. Let N be sufficiently large to ensure that

$$|f_N(x) - f(x)| < \varepsilon/3$$

for all $x \in D$. Now let δ be such that

$$|f_N(x) - f_N(x_0)| < \varepsilon/3$$

for all x such that $|x - x_0| < \delta$. Then for $|x - x_0| < \delta$ we have

$$\begin{aligned}
|f(x) - f(x_0)| &\leq |f(x) - f_N(x) + f_N(x) - f_N(x_0) + f_N(x_0) - f(x_0)| \\
&\leq |f(x) - f_N(x)| + |f_N(x) - f_N(x_0)| + |f_N(x_0) - f(x_0)| \\
&< \varepsilon/3 + \varepsilon/3 + \varepsilon/3 = \varepsilon.
\end{aligned}$$

4.3   Convergence of Fourier series

We know from Theorem 3.3 that for $f \in L_2(-L, L)$, the Fourier series of f converges in the mean to f. In other words

$$\lim_{N \to \infty} \int_{-L}^{L} [S_N(x) - f(x)]^2 \, dx = 0,$$

where

$$P_N f(x) \equiv S_N(x) = a_0 + \sum_{n=1}^{N} \left(a_n \cos n\frac{\pi x}{L} + b_n \sin n\frac{\pi x}{L} \right).$$

(The coefficients a_n and b_n are, of course, the Fourier coefficients.) In practice we generally need to know more and are interested in conditions ensuring point-wise or uniform convergence of the series. In this section are some of the most important results regarding convergence of Fourier series.

This first result is an immediate consequence of Proposition 4.3.

Proposition 4.9 (Bessel's Inequality)

$$a_0^2 + \sum_{n=1}^{\infty} (a_n^2 + b_n^2) \leq \frac{1}{L} \int_{-L}^{L} (f(t))^2 \, dt.$$

The next one follows directly from Corollary 4.5.

Proposition 4.10 (Riemann's Lemma)

$$\lim_{n\to\infty} \frac{1}{L}\int_{-L}^{L} f(t)\cos n\frac{\pi t}{L}\,dt = 0 \ \text{ and } \ \lim_{n\to\infty} \frac{1}{L}\int_{-L}^{L} f(t)\sin n\frac{\pi t}{L}\,dt = 0.$$

Definitions A function f defined on the reals is piecewise continuous if it is continuous except for at most a finite number of jump discontinuities on any finite interval. A function f is piecewise smooth if it is piecewise continuous and on every finite interval has a piecewise continuous derivative except at a finite number of points. Recall that a jump discontinuity of f at a point a is a point at which f is not continuous, but both the one-sided limits $f(a-)$ and $f(a+)$ exist.

Example 4.11 a) The function from Example 4.1, the periodic extension $\tilde{f}$ of the function $f(x) = x$, for $-\pi \le x \le \pi$, is piecewise smooth.
  b) The function f defined by

$$f(x) = \begin{cases} x\sin\frac{1}{x} & x\neq 0 \\ 0 & x=0 \end{cases}$$

is piecewise continuous (in fact, continuous), but is not piecewise smooth. There is no derivative at $x = 0$ and for $x \neq 0$, we have

$$f'(x) = \sin\frac{1}{x} - \frac{1}{x}\cos\frac{1}{x},$$

which has no one-sided limits at $x = 0$. The derivative is thus not piecewise continuous.

We shall now cite the two main theorems on the convergence of Fourier series. The proofs are omitted and may be found, e.g., in [4].

Theorem 4.12 *If f is a piecewise smooth periodic function, then the Fourier series of f converges pointwise to the function*

$$\tilde{f}(x) = \frac{f(x+)+f(x-)}{2}.$$

In case f is continuous, $f(x+) = f(x-)$ for every x and so the Fourier series converges to f. We can, however, in this case say a lot more.

Theorem 4.13 *If f is a continuous piecewise smooth periodic function, then the Fourier series of f converges uniformly to f.*

Since the uniform limit of a sequence of continuous functions is continuous (Theorem 4.8), we know if f is not continuous, then the convergence cannot possibly be uniform.

Example 4.14 a) Let us find the Fourier series for the function $f(x) = x$ on the interval $[-\pi, \pi]$

$$a_n = \frac{1}{\pi} \int_{-\pi}^{\pi} f(t) \cos nt \, dt = \frac{1}{\pi} \int_{-\pi}^{\pi} t \cos nt \, dt = 0,$$

$$b_n = \frac{1}{\pi} \int_{-\pi}^{\pi} f(t) \sin nt \, dt = \frac{1}{\pi} \int_{-\pi}^{\pi} t \sin nt \, dt = \frac{1}{n^2} \left(\sin nt - nt \cos nt \right) \Big|_{-\pi}^{\pi}$$

$$= -\frac{2}{n} \cos n\pi = (-1)^{n+1} \frac{2}{n}.$$

The Fourier series is

$$f \sim 2 \sum_{n=1}^{\infty} \frac{(-1)^{n+1}}{n} \sin nx.$$

Now, what does the sum of this series look like? We simply apply Theorem 4.13 to the periodic extension $\tilde{f}$ of f found in Example 4.1. At discontinuities of the extension (odd multiples of π), the limit is simply 0.

b) We shall find the Fourier series of the function $g(x) = |x|$ on the interval $[-1, 1]$.

$$a_0 = \int_{-1}^{1} |t| \, dt = \frac{1}{2}$$

$$a_n = \int_{-1}^{1} |t| \cos n\pi t \, dt = 2 \int_{0}^{1} t \cos n\pi t \, dt = 2 \frac{-1 + (-1)^n}{n^2 \pi^2} \qquad \text{for } n \geq 1;$$

$$b_n = \int_{-1}^{1} |t| \sin n\pi t \, dt = 0.$$

Life can be made a bit simpler by noting that for $n \geq 1$

$$a_n = \begin{cases} 0 & \text{if } n \text{ even} \\ \frac{-4}{n^2 \pi^2} & \text{if } n \text{ odd}. \end{cases}$$

Then letting $n = 2k + 1$, we have the Fourier series

$$g \sim \frac{1}{2} - \frac{4}{\pi^2} \sum_{k=0}^{\infty} \frac{1}{(2k+1)^2} \cos(2k+1)\pi x.$$

Now, where does this series converge and what does the sum look like? Again, we simply look at the periodic extension $\tilde{g}$ of g. A picture is shown in Fig. 4.4.

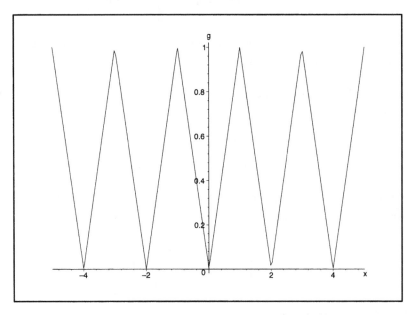

Figure 4.4: Periodic extension of $g(x) = |x|$, $x \in [-1, 1]$.

$\tilde{g}$ is continuous and piecewise smooth so that the Fourier series of g converges uniformly to g.

4.4 Cosine and sine series

We know from the Sturm–Liouville theory that

$$\left\{ 1, \cos\frac{\pi}{L}x, \cos 2\frac{\pi}{L}x, \ldots, \cos n\frac{\pi}{L}x, \ldots \right\} \text{ and}$$

$$\left\{ \sin\frac{\pi}{L}x, \sin 2\frac{\pi}{L}x, \ldots, \sin n\frac{\pi}{L}x, \ldots \right\}$$

are both orthogonal sets in the space $L_2(0, L)$ and that for $f \in L_2(0, L)$, the series

$$A_0 + \sum_{n=1}^{\infty} A_n \cos n\frac{\pi}{L}x, \quad \text{where } A_n = \begin{cases} \frac{1}{L}\int_0^L f(x)dx, & n = 0 \\ \\ \frac{2}{L}\int\limits_0^L f(x)\cos n\frac{\pi}{L}x\,dx, & n \geq 1 \end{cases}$$

converges in the mean to $f(x)$ as does the series

$$\sum_{n=1}^{\infty} B_n \sin n \frac{\pi}{L} x, \quad \text{where } B_n = \frac{2}{L} \int_0^L f(x) \sin n \frac{\pi}{L} x \, dx.$$

These are called, respectively, the Fourier cosine series of f and the Fourier sine series of f. These are very important in the applications to follow.

Definitions Let f be a function defined on the interval $0 \le x \le L$ and let $\widetilde{f}$ be defined on $-L \le x \le L$ by

$$\widetilde{f}(x) = \begin{cases} f(-x) & -L \le x < 0 \\ f(x) & 0 \le x \le L. \end{cases}$$

The periodic extension of $\widetilde{f}$ to the entire real line is called the even periodic extension of f. The periodic extension of f given by

$$\widehat{f}(x) = \begin{cases} -f(-x) & -L \le x < 0 \\ f(x) & 0 < x \le L \\ 0 & x = 0 \end{cases}$$

is called the odd periodic extension of f.

For f defined on $0 \le x \le L$ observe that

$$\int_{-L}^{L} \widetilde{f}(x) \cos n \frac{\pi}{L} x \, dx = 2 \int_0^L f(x) \cos n \frac{\pi}{L} x \, dx; \quad \int_{-L}^{L} \widetilde{f}(x) \sin n \frac{\pi}{L} x \, dx = 0, \quad \text{and}$$

$$\int_{-L}^{L} \widehat{f}(x) \sin n \frac{\pi}{L} x \, dx = 2 \int_0^L f(x) \sin n \frac{\pi}{L} x \, dx; \quad \int_{-L}^{L} \widehat{f}(x) \sin n \frac{\pi}{L} x \, dx = 0.$$

It is clear from these that the Fourier cosine series of f is simply the Fourier series of the even periodic extension $\widetilde{f}$ and the Fourier sine series of f is the Fourier series of the odd periodic extension $\widehat{f}$.

Example 4.15 Let f be defined by $f(x) = 1 - x$ for $0 \le x \le 1$.
a) First, we find the cosine series of f.

$$A_0 = \int_0^1 (1 - x) dx \text{ and for } n \ge 1$$

and we have

$$A_n = 2 \int_0^1 (1 - x) \cos n\pi x \, dx = \frac{2}{\pi^2} \frac{1 - (-1)^n}{n^2}.$$

The cosine series is thus

$$\tilde{f} \sim \frac{1}{2} + \frac{2}{\pi^2} \sum_{n=1}^{\infty} \frac{1 - (-1)^n}{n^2} \cos n\pi x;$$

and the graph of the sum of this series is given in Fig. 4.5.

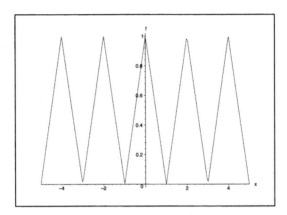

Figure 4.5: Plot of the Fourier cosine series of $f(x) = 1 - x$, $x \in [0, 1]$.

b) Now we find for the sine series of f.

$$B_n = 2 \int_0^1 (1 - x) \sin n\pi x \, dx = \frac{2}{n\pi},$$

and so the sine series is

$$\hat{f} \sim \frac{2}{\pi} \sum_{n=1}^{\infty} \frac{1}{n} \sin n\pi x.$$

The limit of this one at $x \neq 2k$, $k = 0, \pm 1, \ldots$, is shown in Fig. 4.6.

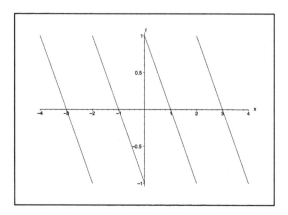

Figure 4.6: Plot of the Fourier sine series of $f(x) = 1 - x$, $x \in [0,1]$.

Observe that both the cosine and the sine series converge to f on the interval $0 < x < 1$, but the cosine series converges uniformly, while the sine series does not.

4.5   Operations on Fourier series

It is clear that if f has Fourier coefficients a_n and b_n, then the function cf, where c is a constant, has Fourier coefficients ca_n and cb_n. Similarly, the coefficients of the sum of two functions are the sums of the corresponding coefficients of the two functions. Here we consider more interesting operations on Fourier series of functions. First, we give the integration of series.

Theorem 4.16 *Suppose f is piecewise continuous and periodic with period $2L$. Then for all x it is true that*

$$\int_c^x f(t)dt = \int_c^x \left[a_0 + \sum_{n=1}^{\infty} \left(a_n \cos \frac{n\pi t}{L} + b_n \sin \frac{n\pi t}{L} \right) \right] dt = a_0(x - c)$$

$$+ \sum_{n=1}^{\infty} \frac{L}{n\pi} \left[a_n \left(\sin \frac{n\pi x}{L} - \sin \frac{n\pi c}{L} \right) - b_n \left(\cos \frac{n\pi x}{L} - \cos \frac{n\pi c}{L} \right) \right]$$

where

$$f \sim a_0 + \sum_{n=1}^{\infty} \left(a_n \cos n \frac{\pi x}{L} + b_n \sin n \frac{\pi x}{L} \right)$$

is the Fourier series of f.

Proof. Let F be defined by

$$F(x) = \int_c^x [f(t) - a] \, dt.$$

Then F is continuous, piecewise smooth and $F'(x) = f(x)$ at the points of continuity of f. Moreover, F is periodic with period $2L$

$$F(x + 2L) = \int_c^{x+2L} [f(t) - a_0]dt$$

$$= \int_c^x [f(t) - a_0]dt + \int_x^{x+2L} [f(t) - a_0]dt = F(x)$$

because

$$\int_x^{x+2L} f(t)dt = \int_{-L}^L f(t)dt = 2La_0.$$

Thus the Fourier series of F converges uniformly to $F(x)$ for all x

$$F(x) = A_0 + \sum_{n=1}^\infty \left(A_n \cos \frac{n\pi x}{L} + B_n \sin \frac{n\pi x}{L} \right)$$

where

$$A_n = \begin{cases} \frac{1}{2L} \int_{-L}^L F(x)dx, & n = 0 \\ \frac{1}{L} \int_{-L}^L F(x) \cos \frac{n\pi x}{L} dx, & n \geq 1 \end{cases}, \quad B_n = \frac{1}{L} \int_{-L}^L F(x) \sin \frac{n\pi x}{L} dx.$$

For $n \neq 0$, integrating by parts gives

$$A_n = -\frac{L}{n\pi} b_n \quad \text{and} \quad B_n = \frac{L}{n\pi} a_n.$$

Hence

$$F(x) = A_0 + \sum_{n=1}^\infty \frac{1}{n} \left(-b_n \cos n \frac{\pi x}{L} + a_n \sin n \frac{\pi x}{L} \right).$$

Now by the definition of a_0

$$F(c + 2L) = \int_c^{c+2L} [f(t) - a_0]dt = 0.$$

This gives

$$A_0 = \sum_{n=1}^\infty \frac{L}{n\pi} \left(b_n \cos \frac{n\pi(c + 2L)}{L} - a_n \sin \frac{n\pi(c + 2L)}{L} \right)$$

so that

$$F(x) = \int_c^x f(t)dt - a_0(x - c)$$

$$= \sum_{n=1}^\infty \frac{L}{n\pi} \left[a_n \left(\sin \frac{n\pi x}{L} - \sin \frac{n\pi c}{L} \right) - b_n \left(\cos \frac{n\pi x}{L} - \cos \frac{n\pi c}{L} \right) \right].$$

Example 4.17 a) In Example 4.14a we found the Fourier series for $f(x) = x$, $-\pi < x < \pi$

$$f \sim 2 \sum_{n=1}^{\infty} \frac{(-1)^{n+1}}{n} \sin nx.$$

Thus for $-\pi \leq x \leq \pi$, it is true that

$$x^2 = 2 \int_0^x t \, dt = \sum_{n=1}^{\infty} \frac{4(-1)^n}{n^2} (\cos nx - 1).$$

It follows from the Fourier series for the 2π-periodic extension of $f(x) = x^2$

$$f \sim \frac{\pi^2}{3} + \sum_{n=1}^{\infty} \frac{4(-1)^n}{n^2} \cos nx,$$

that at $x = 0$

$$\sum_{n=1}^{\infty} \frac{4(-1)^n}{n^2} = -\frac{\pi^2}{3}$$

and hence that

$$x^2 = 2 \int_0^x t \, dt$$

is the Fourier series of $f(x)$.

b) From Example 4.14b, the Fourier series of $f(x) = |x|$, $-1 < x < 1$, is

$$f \sim \frac{1}{2} - \frac{4}{\pi^2} \sum_{k=0}^{\infty} \frac{1}{(2k+1)^2} \cos(2k+1)\pi x.$$

Thus for $-1 \leq x \leq 1$ it is true that

$$\int_{-1}^x |t| \, dt = \frac{1}{2} \begin{cases} 1 - x^2 & x < 0 \\ 1 + x^2 & x \geq 0 \end{cases} = \frac{x+1}{2} - \frac{4}{\pi^3} \sum_{k=0}^{\infty} \frac{1}{(2k+1)^3} \sin(2k+1)\pi x.$$

Observe that Theorem 4.16 does not require that the Fourier series of f converge. Note also that the result of integrating a Fourier series is not in general a Fourier series. In Example 4.17a, the result of integrating the Fourier series of the given function is itself a Fourier series, while in Example 4.17b, the result of the integration is *not* a Fourier series.

We consider next the differentiation of a Fourier series.

Theorem 4.18 *Suppose f is continuous, periodic with period $2L$, and has a piecewise smooth derivative f'. If f' is continuous at x, then*

$$f'(x) = \sum_{n=1}^{\infty} \frac{n\pi}{L} \left(-a_n \sin n\frac{\pi x}{L} + b_n \cos n\frac{\pi x}{L} \right),$$

where a_n and b_n are the Fourier coefficients of f.

Proof. Since f' is continuous at x, applying Theorem 4.13 to f' gives

$$f'(x) = A_0 + \sum_{n=1}^{\infty} \left(A_n \cos n\frac{\pi x}{L} + B_n \sin n\frac{\pi x}{L} \right),$$

where

$$A_n = \frac{1}{L} \int_{-L}^{L} f'(x) \cos n\frac{\pi x}{L} \, dx \quad \text{and} \quad B_n = \frac{1}{L} \int_{-L}^{L} f'(x) \cos n\frac{\pi x}{L} \, dx, \qquad n \geq 1.$$

For $n \neq 0$, integrating by parts now yields

$$A_n = f(x) \cos n\frac{\pi x}{L} \Big|_{-L}^{L} + \frac{n\pi}{L} \int_{-L}^{L} f(x) \sin n\frac{\pi x}{L} dx = \frac{n\pi}{L} b_n.$$

Similarly obtained is

$$B_n = -\frac{n\pi}{L} a_n.$$

Finally

$$A_0 = \int_{-L}^{L} f'(x) dx = f(L) - f(-L) = 0.$$

Substitution of these values for A_n and B_n into the series expression for $f'(x)$ gives the desired result.

Example 4.19 In Example 4.14b, we saw that the periodic extension of the function f defined by $f(x) = |x|$ on the interval $-1 \leq x \leq 1$ satisfies the hypotheses of Theorem 4.13. Thus, since

$$\frac{1}{2} - \frac{4}{\pi^2} \sum_{k=0}^{\infty} \frac{1}{(2k+1)^2} \cos(2k+1)\pi x$$

is the Fourier series of f, we know that

$$\widetilde{f}'(x) = \frac{4}{\pi} \sum_{k=0}^{\infty} \frac{1}{(2k+1)} \sin(2k+1)\pi x$$

for all $x \neq 0, \pm 1, \pm 2, \ldots$, where $\widetilde{f}$ is the periodic extension of f.

Example 4.20 Let u be defined on the interval $[0, L]$, and suppose $u(0) = u(L) = 0$. Suppose further that u is differentiable and $u' \in L_2(0, L)$. We shall show that

$$\int_0^L u^2(x) dx \leq \left(\frac{L}{\pi} \right)^2 \int_0^L (u'(x))^2 \, dx. \tag{4.2}$$

Begin by observing that

$$u(x) = \sum_{n=1}^{\infty} B_n \sin n \frac{\pi}{L} x, \quad \text{where } B_n = \frac{2}{L} \int_0^L u(x) \sin n \frac{\pi}{L} x \, dx.$$

Thus Corollary 4.5 tells us that

$$\int_0^L u^2(x) dx = \frac{L}{2} \sum_{n=1}^{\infty} B_n^2.$$

Next, we know that

$$u'(x) = \sum_{n=1}^{\infty} \frac{n\pi}{L} B_n \cos n \frac{\pi}{L} x,$$

and so

$$\int_0^L \left(u'(x) \right)^2 dx = \frac{L}{2} \sum_{n=1}^{\infty} \left(\frac{n\pi}{L} \right)^2 B_n^2.$$

Then

$$\left(\frac{\pi}{L} \right)^2 \int_0^L u^2(x) dx = \left(\frac{\pi}{L} \right)^2 \frac{L}{2} \sum_{n=1}^{\infty} B_n^2 \le \left(\frac{\pi}{L} \right)^2 \frac{L}{2} \sum_{n=1}^{\infty} n^2 B_n^2$$

$$= \frac{L}{2} \sum_{n=1}^{\infty} \left(\frac{n\pi}{L} \right)^2 B_n^2 = \int_0^L \left(u'(x) \right)^2 dx.$$

Note that we have strict inequality unless $B_n = 0$ for all $n \ge 2$. This inequality bounding the mean square value of u by that of its derivative, and its multidimensional analogue, is known as a Poincaré inequality and plays an important role in the analysis of differential equations and their numerical solution. We shall employ such an inequality in Section 8.3.

4.6   Partial sums of the Fourier series and the Gibbs phenomenon

In applications it is important to have a series with rapidly decreasing coefficients in order that the sum of the first few terms in the series suffice to give an accurate approximation of the limit of the series. Generally, the smoother a function is the more rapidly its Fourier coefficients go to zero. Specifically, we have the following.

Theorem 4.21 *Suppose f is continuous and periodic with period $2L$. Suppose further that f has M derivatives and the Mth derivative is piecewise continuous. Then*

$$\lim_{n\to\infty} n^M a_n = \lim_{n\to\infty} n^M b_n = 0,$$

where a_n and b_n are the Fourier coefficients of f.

Proof. We simply apply Theorem 4.18 M times to get the coefficients in the Fourier series for $f^{(M)}(x)$

$$\frac{n^M}{L^M} a_n \quad \text{and} \quad \frac{n^M}{L^M} b_n.$$

Thus

$$\lim_{n\to\infty} \frac{n^M}{L^M} a_n = \lim_{n\to\infty} \frac{n^M}{L^M} b_n = \frac{1}{L^M} \lim_{n\to\infty} n^M a_n = \frac{1}{L^M} \lim_{n\to\infty} n^M b_n = 0.$$

Example 4.22 a) In Example 4.14b, we found that the Fourier series of $f(x) = |x|$, $-1 \leq x \leq 1$, is

$$f \sim \frac{1}{2} - \frac{4}{\pi^2} \sum_{k=0}^{\infty} \frac{1}{(2k+1)^2} \cos(2k+1)\pi x,$$

and we see that $\lim_{k\to\infty} k^1 a_k = 0$, reflecting the fact that the periodic extension of f has a piecewise smooth first derivative.

b) The Fourier series of f given by

$$f(x) = \begin{cases} (x+1/2)^2 - 1/4 & -1 \leq x < 0 \\ -(-x+1/2)^2 + 1/4 & 0 \leq x \leq 1 \end{cases}$$

is easily found

$$f \sim \frac{-4}{\pi^3} \sum_{n=1}^{\infty} \frac{[(-1)^n - 1]}{n^3} \sin(n\pi x).$$

Here we see that $\lim_{n\to\infty} n^2 b_n = 0$, reflecting that the periodic extension of f has a piecewise continuous second derivative.

We now consider another problem arising in the approximation of a Fourier series by partial sums. Let us look at a simple example. The Fourier series of the function f defined by

$$f(x) = \begin{cases} -\pi - x & -\pi < x \leq 0 \\ \pi - x & 0 < x < \pi \end{cases}$$

is easily found to be

$$f \backsim 2 \sum_{n=1}^{\infty} \frac{\sin nx}{n}.$$

Now, look at a picture in Fig. 4.7: first, a graph of the sum of first 50 terms of
the series.

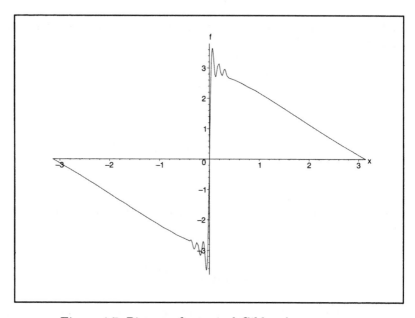

Figure 4.7: Picture of a typical Gibbs phenomenon.

This approximation looks fairly nice except near $x = 0$, where f fails to be
continuous. There is, of course, inevitably a "problem" at a point where f is not
continuous since a partial sum of the Fourier series is necessarily continuous;
but the situation is more complicated than that. We know from Theorem 4.12
that we have pointwise convergence

$$\lim_{N \to \infty} 2 \sum_{n=1}^{N} \frac{\sin nx}{n} = \pi - x \qquad \text{for any } x \in (0, \pi),$$

but as we saw in Example 4.6a, pointwise convergence does not imply uniform
convergence. As we shall prove, the oscillations near $x = 0$ shown in Fig. 4.7
will always be present in the interval $(0, x)$, and their magnitude will remain
constant. For any given $x > 0$ we can squeeze the oscillations into the interval
$(0, x)$ by taking sufficiently many terms in our partial sum, but we cannot elim-
inate them. There is an "overshoot" at the place where f fails to be continuous.
The overshoot and the oscillations around it are called a Gibbs phenomenon

in honor of the physicist J. Willard Gibbs (1839–1903). An explanation of this phenomenon is based on the following result.

Proposition 4.23 *Let g be the discontinuous function given by*

$$g(x) = \begin{cases} -\pi - x & -\pi < x \leq 0 \\ \pi - x & 0 < x < \pi \end{cases}$$

and let

$$S_N(x) = 2 \sum_{n=1}^{N} \frac{\sin nx}{n}$$

be the sum of the first N terms of the Fourier series of g. Then

$$\lim_{N \to \infty} S_N\left(\frac{A}{N}\right) = 2 \int_0^A \frac{\sin t}{t} \, dt.$$

Proof. For given N set $t_n = \frac{nA}{N}$, $\Delta t = \frac{A}{N}$, then

$$S_N\left(\frac{A}{N}\right) \equiv 2 \sum_{n=1}^{N} \frac{\sin t_n}{t_n} \Delta t$$

is a Riemann sum approximation of the integral $2 \int_0^A \frac{\sin t}{t} \, dt$. Since the function $f(t) = \frac{\sin t}{t}$ is continuously differentiable on $[0, A]$, we know that

$$\left| 2 \int_0^A f(t) dt - S_N\left(\frac{A}{N}\right) \right| \leq K \Delta t$$

where $K = \max_{t \in [0,A]} |f'(t)|$. Hence

$$S_N\left(\frac{A}{N}\right) = 2 \int_0^A f(t) dt + R(\Delta t)$$

where

$$|R(\Delta t)| \leq K \Delta t.$$

Corollary 4.24

$$\max_A \left[\lim_{N \to \infty} S_N\left(\frac{A}{N}\right) \right] = \lim_{N \to \infty} S_N\left(\frac{\pi}{N}\right) = 2 \int_0^\pi \frac{\sin t}{t} \, dt = 3.7039.$$

Proof. It is straightforward to verify that the function

$$h(A) = \int_0^A \frac{\sin t}{t} \, dt$$

has relative extrema at $A = n\pi$, that $A = \pi$ corresponds to an absolute maximum on $[0, \infty)$, and that subsequent extrema yield monotonely increasing relative minima and decreasing maxima. It follows that

$$\lim_{N \to \infty} S_N \left(\frac{\pi}{N} \right) \geq \lim_{N \to \infty} S_N \left(\frac{A}{N} \right) \qquad \text{for all } A \in [0, \infty).$$

The "overshoot" is thus

$$\lim_{N \to \infty} S_N \left(\frac{\pi}{N} \right) - g(0+) = 2 \int_0^\pi \frac{\sin t}{t} \, dt - \pi.$$

We can relate the overshoot to the magnitude of the jump discontinuity of g by writing

$$\frac{g(0+) - g(0-)}{2\pi} \left[2 \int_0^\pi \frac{\sin t}{t} \, dt - \pi \right]$$

$$= [g(0+) - g(0-)] \left[\frac{1}{\pi} \int_0^\pi \frac{\sin t}{t} \, dt - \frac{1}{2} \right]$$

$$\cong 8.949 \times 10^{-2} [g(0+) - g(0-)].$$

Hence the overshoot for this function amounts to almost 9 percent of the magnitude of the jump discontinuity of g.

This behavior is not peculiar to this particular function g but occurs at a jump discontinuity of any function f. For example, we have the following theorem.

Theorem 4.25 *Suppose f is piecewise smooth and continuous everywhere on the interval $[-\pi, \pi]$ except at $x = 0$. Let $S_N(x)$ be the Nth partial sum of the Fourier series for f. Then*

$$\lim_{N \to \infty} \left[S_N \left(\frac{\pi}{N} \right) - f \left(\frac{\pi}{N} \right) \right] = [f(0+) - f(0-)] \left(\frac{1}{\pi} \int_0^\pi \frac{\sin t}{t} \, dt - \frac{1}{2} \right).$$

Proof. Let Ψ be defined by

$$\Psi(x) = f(x) - \frac{1}{2} [f(0+) + f(0-)] - \frac{1}{2\pi} [f(0+) - f(0-)] g(x),$$

where g is the function defined in Proposition 4.23. Now

$$\Psi(0+) = f(0+) - \frac{1}{2} [f(0+) + f(0-)] - \frac{1}{2\pi} [f(0+) - f(0-)] g(0+)$$

$$= f(0+) - \frac{1}{2} [f(0+) + f(0-)] - \frac{1}{2\pi} [f(0+) - f(0-)] \pi$$

$$= 0.$$

Similarly, we have $\Psi(0-) = 0$, and so Ψ is continuous everywhere on the given interval. The sequence of partial sums (Ψ_N) of the Fourier series for Ψ thus converges uniformly to Ψ on an interval about 0. Hence

$$\lim_{N \to \infty} \left[\Psi_N \left(\frac{\pi}{N} \right) - \Psi \left(\frac{\pi}{N} \right) \right] = 0.$$

Now

$$\Psi_N \left(\frac{\pi}{N} \right) - \Psi \left(\frac{\pi}{N} \right) = S_N \left(\frac{\pi}{N} \right) - f \left(\frac{\pi}{N} \right)$$
$$- \frac{1}{2\pi} [f(0+) - f(0-)] \left(g_n \left(\frac{\pi}{N} \right) - g \left(\frac{\pi}{N} \right) \right),$$

and so

$$\lim_{N \to \infty} \left\{ S_N \left(\frac{\pi}{N} \right) - f(\frac{\pi}{N}) - \frac{1}{2\pi} [f(0+) - f(0-)] \left(g_n \left(\frac{\pi}{N} \right) - g \left(\frac{\pi}{N} \right) \right) \right\} = 0.$$

Hence

$$\lim_{N \to \infty} \left[S_N \left(\frac{\pi}{N} \right) - f \left(\frac{\pi}{N} \right) \right]$$
$$= \frac{1}{2\pi} [f(0+) - f(0-)] \left[2\pi \left(\frac{1}{\pi} \int_0^\pi \frac{\sin t}{t} \, dt - \frac{1}{2} \right) \right],$$

and we are done.

A Gibbs phenomenon can be expected in the approximation of a function or its derivative with eigenfunctions of a Sturm–Liouville problem whenever the function does not satisfy the boundary conditions of the eigenfunctions. For example, the Fourier cosine series for the function $f(x) = x$ will converge uniformly, but its derivative is the sine series of $f'(x) = 1$ with its Gibbs phenomenon. As a further illustration, let us consider the projection of functions into the span of the eigenfunctions of the Sturm–Liouville problem

$$\phi''(x) = -\lambda^2 \phi(x) \tag{4.3}$$

$$\phi(0) = \phi'(0), \qquad \phi(1) = 0.$$

The eigenfunctions are

$$\phi_n(x) = \sin \lambda_n (1 - x)$$

where λ_n is the nth positive root of

$$f(\lambda) \equiv \sin \lambda + \lambda \cos \lambda = 0.$$

These roots are easy to find numerically. When we project the functions $f_1(x) = 1 - x^2$, $f_2(x) = x - x^2$, and $f_3(x) = f_1(x) + f_2(x)$ onto span $\{\phi_1, \ldots, \phi_N\}$ we

observe uniform convergence of $P_N f_i$ to f_i, $i = 1, 2, 3$. Neither f_1 nor f_2 satisfies
the boundary condition of ϕ at $x = 0$, but f_3 does. Fig. 4.8 shows plots of
$(P_N f_i)'(x)$ for $N = 70$ and $i = 1, 2, 3$. There are pronounced Gibbs effects for
$i = 1, 2$ but which cancel to give (what appears to be) uniform convergence of
$(P_N f_3)'(x)$ to $f_3'(x)$.

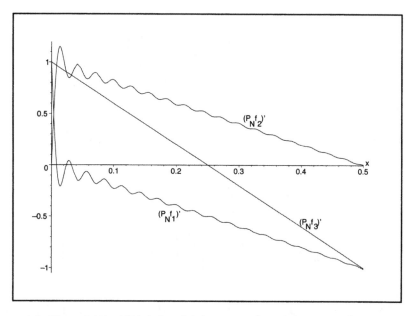

Figure 4.8: Plot of $(P_N f_i)'(x)$ for $f_1(x) = 1 - x^2$, $f_2(x) = x - x^2$, and $f_3(x) =$
$f_1(x) + f_2(x)$ and $N = 70$.

A Web search with the keywords "Gibbs phenomenon" reveals that the
phenomenon is observed whenever discontinuous functions are projected into
the span of orthogonal function such as trigonometric, Bessel, Legendre, and
Chebychev functions. Moreover, it is known that the Gibbs phenomenon can be
canceled by suitably modifying the expansion [11]. For the eigenfunction expansions of the subsequent chapters such changes cannot be carried out because the
orthogonal functions are obtained as solutions of a differential equation. We are
stuck with the Gibbs phenomenon whenever the data to be approximated are
discontinuous. On the other hand, we should be careful to avoid approximations
that introduce discontinuities into the data which are not present in the original
problem. Whenever possible we should work with data which belong to the subspace defined by the eigenfunctions used for their approximation. This concern
leads to the preconditioning of elliptic boundary value problems discussed in
Chapter 8.

Exercises

4.1) Give an example of a sequence $\{f_n\}$ in $L_2(0,1)$ that converges uniformly to a function f such that

$$\lim_{n\to\infty} \int_0^1 f_n(x)dx \neq \int_0^1 f(x)dx,$$

or explain why there can be no such example.

4.2) Give an example of a sequence of functions $\{f_n\}$ in $L_2(0,1)$ differentiable on $(0,1)$ that converges uniformly to a function f that is not differentiable, or explain why there can be no such example.

4.3) Let $f_n(x) = \begin{cases} A, & x = 0 \\ n^2(1/n - x), & 0 < x < 1/n, \quad n > 1 \\ 0, & 1/n \leq x \leq 1 \end{cases}$

where A is a constant.

i) Let $\tilde{f}_n$ be the even extension of f_n to the interval $[-1,1]$. In what sense does $\{\tilde{f}_n\}$ converge as $n \to \infty$? If $\{\tilde{f}_n\}$ converges, what is the limit? Compute $\lim_{n\to\infty} \int_{-1}^1 \tilde{f}_n(x)g(x)dx$ where g is continuous at $x = 0$.

ii) Let $\hat{f}_n$ be the odd extension of f_n to the interval $[-1,1]$. In what sense does $\{\hat{f}_n\}$ converge as $n \to \infty$? If $\{\hat{f}_n\}$ converges, what is the limit? Compute $\lim_{n\to\infty} \int_{-1}^1 \hat{f}_n(x)g(x)dx$ where g is continuous at $x = 0$.

4.4) Let f be a continuous $2L$ periodic function. Show that for any c

$$\int_c^{c+2L} f(x)dx = \int_{-L}^L f(x)dx.$$

4.5) Suppose that f is a differentiable even function on $(-\infty, \infty)$. Show that

i) $f'(x)$ is odd.

ii) $F(x) = \int_0^x f(s)ds$ is odd.

4.6) Find the Fourier series of $f(x) = x^2$, $-\pi \leq x \leq \pi$, and sketch the graph of its limit.

4.7) Let
$$f(x) = \begin{cases} -1 & -\pi \le x < 0 \\ 1 & 0 \le x \le \pi \end{cases}$$

Find the Fourier series of f and sketch the graph of its limit.

4.8) Find the Fourier series of the function f defined on the interval $-1 \le x \le 1$ by
$$f(x) = \begin{cases} 0 & -1 \le x \le 0 \\ x & 0 < x \le 1 \end{cases}$$

and sketch the graph of its limit.

4.9) Show that
$$\sum_{k=0}^{\infty} \frac{1}{(2k+1)^2} = \frac{\pi^2}{8}.$$

Hint: Example 4.5b.

4.10) For each of the following, tell whether or not the Fourier series of the given function converges uniformly, and explain your answers.

i) $f(x) = e^x$, $-1 \le x \le 1$.

ii) $f(x) = x^2$, $-1 \le x \le 1$.

iii) $f(x) = e^{-x^2}$, $-1 \le x \le 1$.

iv) $f(x) = x^3$, $-\pi \le x \le \pi$.

4.11) Let f be defined on the interval $0 \le x \le 2$ by
$$f(x) = \begin{cases} 0 & 0 \le x < 1 \\ 1 & 1 \le x < 2. \end{cases}$$

i) Find the Fourier cosine series of f and sketch the graph of its limit.

ii) Find the Fourier sine series of f and sketch the graph of its limit.

4.12) Let f be defined by $f(x) = \sin x$ for $0 \le x \le \pi$.

i) Find the Fourier sine series of f and sketch the graph of its limit.

ii) Find the Fourier cosine series of f and sketch the graph of its limit.

4.13) Show that the estimate proved in Example 4.20 is sharp for $u(x) = \sin \frac{\pi}{L} x$.

4.14) Let $f(x) = 2 - x$, $x \in [0, 1]$. Without any calculations make a rough sketch over $-4 \leq x \leq 4$ of

   i) Fourier sine series of f,

   ii) Fourier cosine series of f,

   iii) Fourier series of f.

In each case indicate where you would expect a Gibbs phenomenon.

   iv) Find a function $g(x)$ defined on an interval I such that the Fourier series for g converges uniformly to $f(x)$ for $x \in [0, 1]$. Are the interval I and the function g uniquely defined?

4.15) Compute the partial sum S_N of the Fourier sine series for $f(x) = 1/x$ on $[0, 1]$. Does S_N converge as $N \to \infty$?

4.16) Find the orthogonal projection of $f(x) \equiv 1$ into $\text{span}\{\phi_1, \ldots, \phi_N\}$ where ϕ solves equation (4.3). Does $P_N f$ show a Gibbs phenomenon at $x = 0$? Does $(P_N f)'$ show a Gibbs phenomenon at $x = 0$? The eigenvalues will have to be found numerically.

Chapter 5

Eigenfunction Expansions for Equations in Two Independent Variables

Drawing on the Sturm–Liouville eigenvalue theory and the approximation of functions we are now ready to develop the eigenfunction approach to the approximate solution of boundary value problems for partial differential equations. All these problems have the same basic structure. We shall outline the general solution process and then examine the technical differences which arise when it is applied to the heat, wave and Laplace's equation. Specific applications and numerical examples are discussed in subsequent chapters.

We shall consider partial differential equations in two independent variables (x, t) where usually x denotes a space coordinate and t stands for time. However, on occasion, as in potential problems, both variables may denote space coordinates. In this case we tend to choose (x, y) as independent variables.

All problems to be considered are of the form

$$\mathcal{L}u(x, t) = F(x, t) \qquad (5.1)$$

where $\mathcal{L}$ is a linear partial differential operator, possibly with variable coefficients, defined for $x \in (0, L)$ and $t \in (0, T]$ (or $y \in (0, b)$).

Typical examples to be discussed at length in Chapters 6–8 are the heat equation

$$\mathcal{L}u \equiv u_{xx} - u_t = F(x, t),$$

the wave equation

$$\mathcal{L}u \equiv u_{xx} - u_{tt} = F(x, t),$$

and Poisson's equation (also called the potential equation)

$$\mathcal{L}u \equiv u_{xx} + u_{yy} = F(x,y).$$

It is characteristic of our eigenfunction expansion view of separation of variables that in all these equations we admit source terms which are functions of the independent variables.

Equation (5.1) is to be solved for a function u which satisfies linear homogeneous or inhomogeneous boundary conditions at $x = 0$ and $x = L$. Specifically, for our three model problems we expect that u either satisfies the homogeneous periodicity conditions

$$u(0,t) = u(L,t)$$
$$u_x(0,t) = u_x(L,t)$$

or the general inhomogeneous boundary conditions

$$\alpha_1 u_x(0,t) - \alpha_2 u(0,t) = A(t)$$

$$\beta_1 u_x(L,t) + \beta_2 u(L,t) = B(t)$$

for nonnegative parameters α_1, α_2, β_1, β_2 and smooth functions $A(t)$ and $B(t)$.

In addition, u is required to satisfy an initial condition at $t = 0$ or boundary conditions at $y = 0$ and $y = b$. For definiteness, we shall assume that

$$u(x,0) = u_0(x), \tag{5.2}$$

which is typical for the heat equation. Other conditions are discussed when applying the spectral approach to the wave and potential equation.

For linear inhomogeneous boundary conditions it is possible to find a function $v(x,t)$ which satisfies the boundary conditions imposed on u. For example, suppose that the boundary conditions of the problem are

$$u(0,t) = A(t) \tag{5.3}$$
$$u_x(L,t) = h[B(t) - u(L,t)]$$

where A and B are given functions of t and h is a positive constant. If we choose a function $v(x,t)$ of the form

$$v(x,t) = a(t) + b(t)x, \tag{5.4}$$

then it is straightforward to find a and b such that v satisfies (5.3). We need to solve

$$v(0,t) = a(t) = A(t)$$

$$v_x(L,t) = b(t) = h[B(t) - v(L,t)] = h[B(t) - a(t) - b(t)L]$$

so that

$$a(t) = A(t)$$
$$b(t) = \frac{h[B(t) - A(t)]}{1 + hL}.$$

Note that $v(x,t)$ is not unique. We could have chosen, for example

$$v(x,t) = a(t) + b(t)x + c(t)x^2$$

and determined $a(t)$, $b(t)$, and $c(t)$ so that this v satisfies the given boundary conditions. In this case there will be infinitely many solutions. In general, the form of (5.4) for v is the simplest choice and leads to the least amount of computation, provided a and b can be found. If not, a quadratic in x will succeed (see Example 8.2). On special occasions a v structured to the problem must be used (see Example 6.8).

If we now set

$$w(x,t) = u(x,t) - v(x,t),$$

then w will satisfy one of the boundary conditions listed in Table 3.1. For equation (5.3) we would obtain

$$w(0,t) = 0$$
$$w_x(L,t) = -hw(L,t).$$

In other words, the function $w(x,t)$ as a function of x belongs to one of the subspaces M discussed in Chapter 3, and this subspace does not change with t.

Excluded from our discussion are nonlinear boundary conditions like the so-called radiation condition

$$u_x(L,t) = h[B^4(t) - u^4(L,t)]$$

or a reflection condition like

$$u_x(L,t) = -h(t)u(L,t)$$

for a time-dependent function h.

Transforming the problem for u with inhomogeneous boundary conditions into an equivalent problem for w with homogeneous boundary conditions is the first step in applying any form of separation of variables. Once this is done the problem can be restated for w as:

Find a function $w(x,t)$ which satisfies

$$\mathcal{L}w = \mathcal{L}u - \mathcal{L}v = F(x,t) - \mathcal{L}v \equiv G(x,t), \qquad (5.5)$$

which satisfies the corresponding homogeneous boundary conditions at $x = 0$ and $x = L$, and which satisfies the given conditions at $t = 0$ (and, if applicable, at $y = b$), i.e., here

$$w(x,0) = u_0(x) - v(x,0) \equiv w_0(x). \qquad (5.6)$$

We emphasize that G and w_0 are known data functions.

We now make the following two essential assumptions which lie at the heart of any separation of variables method:

I) The partial differential equation (5.5) can be written in the form

$$\mathcal{L}w(x,t) = \mathcal{L}_1(x)w + \mathcal{L}_2(t)w = G(x,t)$$

where

   i) $\mathcal{L}_1$ denotes the terms involving functions of x and derivatives with respect to x,

   ii) $\mathcal{L}_2$ denotes the terms involving functions of t and derivatives with respect to t.

II) The eigenvalue problem

$$\mathcal{L}_1(x)\phi = \mu\phi,$$

subject to one of the boundary conditions of (3.2), has obtainable solutions $\{\mu_n, \phi_n\}_{n=1}^N$. In all of our applications the eigenvalue problem is a Sturm–Liouville eigenvalue problem so that the eigenfunctions are orthogonal in an inner product space M which is determined by $\mathcal{L}_1(x)$ and the boundary conditions.

The computation of an approximate solution of (5.5), (5.6) is now automatic. We define

$$M_N = \mathrm{span}\{\phi_1(x), \ldots, \phi_N(x)\}.$$

We compute the best approximations, i.e., the projections

$$P_N G(x,t) = \sum_{n=1}^{N} \gamma_n(t)\phi_n(x)$$

$$P_N w_0(x) = \sum_{n=1}^{N} \hat{\alpha}_n \phi_n(x)$$

of the space-dependent data functions (treating t as a parameter) and solve the approximating problem

$$\mathcal{L}w = P_N G(x,t) \tag{5.7}$$

$$w(x,0) = P_N w_0(x).$$

We compute an exact solution $w(x,t)$ of (5.7) by assuming that it belongs to M_N for all t. In this case it has to have the form

$$w_N(x,t) = \sum_{n=1}^{N} \alpha_n(t)\phi_n(x).$$

We want $\mathcal{L}w_N = P_N G$ and $w_N(x,0) = P_N w_0(x)$ so the coefficients $\{\alpha_n(t)\}$ must be chosen such that

$$\sum_{n=1}^{N} [\mu_n \alpha_n(t) + \mathcal{L}_2(t)\alpha_n(t) - \gamma_n(t)]\phi_n(x) = 0.$$

Since the eigenfunctions $\{\phi_n(x)\}$ are linearly independent, the term in the bracket must vanish. Hence each coefficient $\alpha_n(t)$ has to satisfy the ordinary differential equation

$$\mu_n \alpha_n(t) + \mathcal{L}_2(t)\alpha_n = \gamma_n(t)$$

and the initial condition

$$\alpha_n(0) = \hat{\alpha}_n.$$

The techniques of ordinary differential equations give us explicit solutions $\{\alpha_n(t)\}$. It may generally be assumed that the problem (5.7) for w is well posed so that the w_N just constructed is the only solution of (5.7).

So far the specific form of $\mathcal{L}_2(t)$ has not entered our discussion. Hence, regardless of whether we solve the heat equation, the wave equation, or Laplace's equation, the solution process always consists of the following steps:

Step 1: Find a function v which satisfies the same boundary conditions at $x = 0$ and $x = L$ as the unknown solution $u(x,t)$.

Step 2: Set $w = u - v$ and write problem (5.5)

$$\mathcal{L}w = G(x,t),$$

the linear homogeneous boundary conditions at $x = 0$ and $x = L$, and the conditions for w at $t = 0$ (or at $y = 0$ and $y = b$).

Step 3: For these boundary conditions solve the eigenvalue problem

$$\mathcal{L}_1(x)\phi_n = \mu_n \phi_n, \qquad n = 1, \ldots, N$$

for the first N eigenvalues and eigenvectors.

Step 4: Project $G(x,t)$ and the initial or boundary conditions at $t = 0$ (or at $y = 0$ and $y = b$) into the span of these N eigenfunctions, treating t as a parameter, to obtain the approximating problem (5.7).

Step 5: Solve for w_N.

Step 6: Accept as an approximation to the solution u of the original problem the computed solution

$$u_N = w_N + v.$$

To illustrate the problem independence of these steps, but also to highlight some of the computational differences in carrying them out for varying initial and boundary conditions we shall discuss in a qualitative sense the solution process for the heat, the wave, and the potential equation.

The eigenfunction method, also known as spectral method, is easiest to apply to the heat equation. Thus let us consider the initial/boundary value problem

$$\mathcal{L}u \equiv u_{xx} - u_t = F(x,t), \qquad x \in (0,L), \quad t > 0$$

$$u(0,t) = A(t), \qquad t > 0$$

$$u_x(L,t) = B(t), \qquad t > 0$$

$$u(x,0) = u_0(x), \qquad x \in [0,L].$$

It models the temperature distribution $u(x,t)$ in a slab of thickness L (or an insulated bar of length L) as a function of position and a scaled time. $F(x,t)$ denotes an internal heat source or sink, and $A(t)$ and $B(t)$ are a prescribed (and generally time-dependent) temperature and flux at the ends of the slab or bar. The initial temperature distribution is $u_0(x)$.

In order to rewrite the problem for homogeneous boundary conditions we choose

$$v(x,t) = A(t) + B(t)x,$$

which satisfies the boundary conditions imposed on $u(x,t)$, and define

$$w(x,t) = u(x,t) - v(x,t).$$

Then

$$\mathcal{L}w = \mathcal{L}u - \mathcal{L}v = F(x,t) + v_t = F(x,t) + (A'(t) + B'(t)x) \equiv G(x,t)$$

$$w(0,t) = 0, \qquad w_x(L,t) = 0$$

$$w(0,t) = u_0(x) - v(x,0) \equiv w_0(x).$$

Here we have assumed that A and B are differentiable. We shall see below that the final result depends only on A and B, not their derivatives.

Since

$$\mathcal{L}w \equiv w_{xx} - w_t,$$

we see that

$$\mathcal{L}_1(x)w = w_{xx}, \qquad \mathcal{L}_2(t)w = w_t.$$

The homogeneous boundary conditions at $x = 0$ and $x = L$ dictate that we solve the eigenvalue problem

$$\mathcal{L}_1(x)\phi \equiv \phi''(x) = \mu\phi(x)$$
$$\phi(0) = \phi'(L) = 0.$$

The eigenvalues and eigenfunctions are available from Table 3.1 as

$$\lambda_n = \frac{\left(\frac{1}{2} + n\right)\pi}{L}, \quad \mu_n = -\lambda_n^2, \quad \phi_n(x) = \sin\lambda_n x, \quad n = 0, 1, \ldots.$$

Because the eigenfunctions are orthogonal in $L_2(0, L)$, we readily can approximate the source term G and the initial condition w_0 in the span M_N of the first $N + 1$ eigenfunctions. $P_N G$ and $P_N w_0$ are the orthogonal projections

$$P_N G(x, t) = \sum_{n=0}^{N} \gamma_n(t)\phi_n(x)$$

$$P_N w_0(x) = \sum_{n=0}^{N} \hat{\alpha}_n \phi_n(x)$$

where

$$\gamma_n(t) = \frac{\langle G(x, t), \phi_n(x)\rangle}{\langle \phi_n(x), \phi_n(x)\rangle} = \frac{2}{L}\left[\langle F(x, t), \phi_n\rangle + A'(t)\langle 1, \phi_n\rangle + B'(t)\langle x, \phi_n\rangle\right]$$

$$\hat{\alpha}_n = \frac{2}{L}\langle w_0(x), \phi_n(x)\rangle,$$

with $\langle f, g\rangle = \int_0^L f(x)g(x)dx$. The solution of the approximating problem can be expressed as

$$w_N(x, t) = \sum_{n=0}^{N} \alpha_n(t)\phi_n(x).$$

Substitution into

$$\mathcal{L}w = G_N(x, t)$$

$$w(x, 0) = P_N w_0(x)$$

and use of the eigenvalue equation show that

$$-\lambda_n^2 \alpha_n(t) - \alpha_n'(t) = \gamma_n(t)$$

$$\alpha_n(0) = \hat{a}_n.$$

The solution of this equation has the form

$$\alpha_n(t) = d_n \alpha_{nc}(t) + \alpha_{np}(t)$$

where d_n is a constant, $\alpha_{nc}(t)$ is a complementary solution of the equation

$$\alpha' + \lambda_n^2 \alpha = 0,$$

and $\alpha_{np}(t)$ is a particular integral of the inhomogeneous equation

$$\alpha' + \lambda_n^2 \alpha = -\gamma_n(t).$$

For $\alpha_{nc}(t)$ we choose

$$\alpha_{nc}(t) = e^{-\lambda_n^2 t}.$$

Our ability to find the particular integral analytically will depend crucially on the form of the source term $\gamma_n(t)$. If it is the product of a real or complex exponential and a polynomial, then the method of undetermined coefficients suggests itself. Otherwise the variation of parameters solution can be used which is

$$\alpha_{np}(t) = -\int_0^t e^{-\lambda_n^2 (t-s)} \gamma_n(s) ds.$$

The solution $\alpha_n(t)$ is then

$$\alpha_n(t) = e^{-\lambda_n^2 t} \hat{a}_n - \int_0^t e^{-\lambda_n^2 (t-s)} \gamma_n(s) ds. \qquad (5.8)$$

The approximation to the solution of the original problem is

$$u_N(x, t) = w_N(x, t) + v(x, t).$$

We note that the derivatives of the boundary data $A(t)$ and $B(t)$ occur only under the integral in (5.8). If we apply integration by parts, then the boundary data need be only integrable. For example

$$\int_0^t e^{-\lambda_n^2 (t-s)} A'(s) ds = A(t) - A(0) e^{-\lambda_n^2 t} - \lambda_n^2 \int_0^t e^{-\lambda_n^2 (t-s)} A(s) ds.$$

Hence the assumption that $A(t)$ and $B(t)$ be differentiable may be dispensed with after applying integration by parts to (5.8) (see also Exercises 5.12 and 5.16).

Let us next consider the vibration of a driven uniform string of length L. The problem to be solved is

$$\mathcal{L}u(x,t) \equiv u_{xx} - \frac{1}{c^2} u_{tt} = F(x,t)$$

$$u(0,t) = A(t), \qquad u(L,t) = B(t)$$

with initial conditions

$$u(0,t) = u_0(x)$$

$$u_t(x,0) = u_1(x).$$

We shall assume that the boundary and initial conditions are consistent and smooth so that the problem has a unique smooth solution.

A simple function satisfying the given boundary conditions is

$$v(x,t) = A(t)\frac{L-x}{L} + B(t)\frac{x}{L}.$$

If we set

$$w(x,t) = u(x,t) - v(x,t),$$

then w satisfies the problem

$$\mathcal{L}w = \mathcal{L}u - \mathcal{L}v = F(x,t) + \frac{1}{c^2}\left[A''(t)\frac{L-x}{L} + B''(t)\frac{x}{L}\right] \equiv G(x,t).$$

$$w(x,0) = u_0(x) - v(x,0) \equiv w_0(x)$$

$$w_t(x,0) = u_1(x) - v_t(x,0) \equiv w_1(x).$$

The associated eigenvalue problem is

$$\phi''(x) = \mu\phi(x)$$

$$\phi(0) = \phi(L) = 0$$

which has the solution

$$\phi_n(x) = \sin \lambda_n x, \quad \mu_n = -\lambda_n^2, \quad \lambda_n = \frac{n\pi}{L}, \qquad n = 1, \ldots.$$

These functions are orthogonal in $L_2(0,L)$. The approximating problem is

$$\mathcal{L}w = P_N G(x,t) = \sum_{n=1}^{N} \gamma_n(t)\phi_n(x)$$

$$w(0,t) = w(L,t) = 0$$

$$w(x,0) = P_N w_0(x) = \sum_{n=1}^{N} \hat{\alpha}_n \phi_n(x)$$

$$w_t(x,0) = P_N w_1(x) = \sum_{n=1}^{N} \hat{\beta}_n \phi_n(x)$$

where

$$\gamma_n(t) = \frac{2}{L}\langle G(x,t), \phi_n \rangle, \quad \hat{\alpha}_n = \frac{2}{L}\langle w_0, \phi_n \rangle, \quad \hat{\beta}_n = \frac{2}{L}\langle w_1, \phi_n \rangle.$$

The solution is

$$w_N(x,t) = \sum_{n=1}^{N} \alpha_n(t)\phi_n(x)$$

where

$$-\lambda_n^2 \alpha_n(t) - \frac{1}{c^2}\alpha_n''(t) = \gamma_n(t)$$

$$\alpha_n(0) = \hat{\alpha}_n$$

$$\alpha_n'(0) = \hat{\beta}_n.$$

This equation has a unique solution of the form

$$\alpha_n(t) = d_n^1 \alpha_{nc}^1(t) + d_n^2 \alpha_{nc}^2(t) + \alpha_{np}(t)$$

where $\alpha_{nc}^1(t)$ and $\alpha_{nc}^2(t)$ are two linearly independent solutions of

$$-\lambda_n^2 \alpha_n(t) - \frac{1}{c^2}\alpha_n''(t) = 0,$$

taken here to be

$$\alpha_{nc}^1(t) = \sin\lambda_n ct$$

$$\alpha_{nc}^2(t) = \cos\lambda_n ct.$$

$\alpha_{np}(t)$ is a particular integral of the equation. Its form will depend on the structure of $\gamma_n(t)$. If possible, the method of undetermined coefficients should be applied; otherwise the method of variation of parameters must be applied which yields the formula

$$\alpha_{np}(t) = f_n^1(t)\alpha_{nc}^1(t) + f_n^2(t)\alpha_{nc}^2(t)$$

where

$$f_n^1(t) = \int_0^t \frac{c^2 \alpha_{nc}^2(s)\gamma_n(s)}{\alpha_{nc}^1(s)\alpha_{nc}'^2(s) - \alpha_{nc}'^1(s)\alpha_{nc}^2(s)}\, ds$$

$$f_n^2(t) = \int_0^t \frac{-c^2 \alpha_{nc}^1(s)\gamma_n(s)}{\alpha_{nc}^1(s)\alpha_{nc}'^2(s) - \alpha_{nc}'^1(s)\alpha_{nc}^2(s)}\, ds.$$

These integrals will simplify because

$$\alpha^1_{nc}(s)\alpha'^2_{nc}(s) - \alpha'^1_{nc}(s)\alpha^2_{nc}(s) = \lambda_n c.$$

The variation of parameters solution for $\alpha_n(t)$ is

$$\alpha_n(t) = (d^1_n + f^1_n(t))\sin\lambda_n ct + (d^2_n + f^2_n(t))\cos\lambda_n ct.$$

The initial conditions require that

$$d^1_n = \frac{\hat{\beta}_n}{\lambda_n c}$$

$$d^2_n = \hat{\alpha}_n$$

because $f^1_n(0) = f^2_n(0) = f'^2_n(0) = 0$. The solution of the approximating problem is

$$u_N(x,t) = \sum_{n=1}^{N} \alpha_n(t)\phi_n(x) + v(x,t).$$

We point out that just about the same equations result if we consider the case of a simply supported uniform vibrating beam; see Example 7.8.

For the last illustration of the general eigenfunction expansion approach we turn to the potential problem

$$\mathcal{L}u \equiv u_{xx} + u_{yy} = F(x,y), \quad (x,y) \in (0,L) \times (0,b),$$

$$u_x(0,y) = A(y), \quad u_x(L,y) = B(y), \quad y \in (0,b),$$

$$u(x,0) = C(x), \quad u(x,b) = D(x), \quad x \in [0,L],$$

which may be interpreted as a steady-state heat flow problem in a rectangular plate with prescribed temperatures and fluxes on the boundary. As before we shall assume that the data are smooth functions.

For an actual calculation it would be simpler to employ eigenfunctions in the y-direction, but for this illustration we shall again choose eigenfunctions in the x-direction. To obtain homogeneous boundary conditions at $x = 0$ and $x = L$ we need to find a $v(x,y)$ satisfying

$$v_x(0,y) = A(y), \quad v_x(L,y) = B(y).$$

An extensive discussion of the proper choice of $v(x,y)$ for making the boundary data homogeneous may be found in Examples 8.1 and 8.2. Here we simply observe that this time we cannot succeed with a function which is linear in x

because its derivative only has one degree of freedom. Instead we shall choose the quadratic in x

$$v(x,y) = A(y)x + \frac{B(y) - A(y)}{2L}x^2.$$

The equivalent problem for

$$w(x,y) = u(x,y) - v(x,y)$$

is

$$\mathcal{L}w = \mathcal{L}u - \mathcal{L}v \equiv G(x,y)$$

with

$$G(x,y) = F(x,y) - \left[\frac{B(y) - A(y)}{L} + A''(y)x + \frac{B''(y) - A''(y)}{2L}x^2\right] \equiv G(x,y),$$

$$w_x(0,y) = w_x(L,y) = 0,$$

$$w(x,0) = C(x) - v(x,0) \equiv g_1(x),$$

$$w(x,b) = D(x) - v(x,b) \equiv g_2(x).$$

The associated eigenvalue problem is

$$\phi''(x) = \mu\phi(x)$$

$$\phi'(0) = \phi'(L) = 0.$$

Its first $N+1$ eigenfunctions and eigenvalues are

$$\lambda_n = \frac{n\pi}{L}, \quad \mu_n = -\lambda_n^2, \quad \phi_n(x) = \cos\lambda_n x, \quad n = 0,\ldots,N,$$

which are orthogonal in $L_2(0,L)$. The approximating problem is

$$\mathcal{L}w \equiv P_N G(x,y) = \sum_{n=0}^{N} \gamma_n(y)\phi_n(x)$$

$$w_x(0,y) = w_x(L,y) = 0$$

$$w(x,0) = P_N g_1(x) = \sum_{n=0}^{N} \hat{\alpha}_n \phi_n(x)$$

$$w(x,b) = P_N g_2(x) = \sum_{n=0}^{N} \hat{\beta}_n \phi_n(x)$$

where, e.g.

$$\gamma_n(y) = \frac{2}{L}\langle G(x,y), \phi_n\rangle.$$

The approximating problem is solved by

$$w_N(x, y) = \sum_{n=0}^{N} \alpha_n(y)\phi_n(x)$$

if $\alpha_n(y)$ is a solution of the two-point boundary value problem

$$-\lambda_n^2 \alpha_n(y) + \alpha_n''(y) = \gamma_n(y)$$

$$\alpha_n(0) = \hat{\alpha}_n, \quad \alpha_n(b) = \hat{\beta}_n.$$

Note that a solution of this boundary value problem is necessarily unique. This observation follows from the maximum principle of Section 1.3. Indeed, the difference $e_n(y)$ of two solutions would solve the problem

$$e_n''(y) - \lambda_n^2 e_n(y) = 0$$

$$e_n(0) = e_n(b) = 0.$$

The second derivative test now rules out an interior positive maximum or negative minimum so that $e_n(y) = 0$. The solution of the differential equation is again

$$\alpha_n(y) = d_n^1 \alpha_{nc}^1(y) + d_n^2 \alpha_{nc}^2(y) + \alpha_{np}(y)$$

where now

$$\alpha_{nc}^1(y) = \sinh \lambda_n y$$

$$\alpha_{nc}^2(y) = \cosh \lambda_n y.$$

The particular integral is the same as given above for the wave equation provided c^2 is replaced by -1 and t by y. Hence

$$\alpha_n(y) = (d_n^1 + f_n^1(y)) \sinh \lambda_n y + (d_n^2 + f_n^2(y)) \cosh \lambda_n y.$$

d_n^1 and d_n^2 must now be determined from the boundary conditions

$$\alpha_n(0) = \hat{\alpha}_n$$

$$\alpha_n(b) = \hat{\beta}_n.$$

The final approximate solution of the potential problem is

$$u_N(x, y) = w_N(x, y) + v(x, y).$$

It is natural to ask how w_N is related to the analytic solution $w(x, t)$ of the original problem. For several of the problems considered below we shall prove the following remarkable result:

$$w_N(x, t) = P_N w(x, t).$$

Hence the computed solution is exactly the projection of the unknown analytic solution. In general one has a fair amount of information from the theory of partial differential equations about the smoothness properties of w. In particular, w is nearly always square integrable. The general Sturm–Liouville theory can then be invoked to conclude that, at least in the mean square sense, w_N converges to w as $N \to \infty$. This implies that when our finite sums are replaced by infinite series, then the resulting function is, in a formal sense, the analytic solution $w(x,t)$. Some quantitative estimates for the quality of the approximation can be found for specific problems as outlined in the chapters to follow.

Exercises

5.1) Apply the solution process of this chapter to find an approximate solution of the problem

$$\mathcal{L}u \equiv u'' = 1 + x$$

$$u'(0) = 1, \qquad u(1) = 0,$$

i.e.

    i) What do you choose for v?

    ii) What is the problem satisfied by $w = u - v$?

    iii) What is the associated eigenvalue problem?

    iv) What is the approximating problem?

    v) Solve for w_N.

    vi) Find the analytic solution w.

    vii) Show that $w_N = P_N w$.

    viii) What is the approximate solution u_N of the original problem?

    ix) Compare u_N with the analytic solution u as $N \to \infty$.

5.2) Apply the solution process of this chapter to find an approximate solution of the problem

$$\mathcal{L}u \equiv u'' - u = e^x$$

$$u(0) = 1, \qquad u'(2) = -1.$$

Follow all the steps detailed in problem 5.1.

5.3) Show that the problem

$$\mathcal{L}u \equiv u'' = 1$$

$$u'(0) = u'(1) = 0$$

has no solution.

Show that the solution process of this chapter breaks down when we try to find an approximate solution.

5.4) Show that the problem

$$\mathcal{L}u \equiv u'' = 1$$

$$u'(0) = 0, \qquad u'(1) = 1$$

has infinitely many solutions. Find an approximate eigenfunction solution when the boundary conditions are zeroed out with

$$v(x) = \frac{x^K}{K} \quad \text{for } K > 2.$$

5.5) Apply the solution process of this chapter to find an approximate solution of the problem

$$\mathcal{L}u \equiv u_{xx} - u_t = 1$$

$$u(0, t) = 0, \qquad u(\pi, t) = 0$$

$$u(x, 0) = 0.$$

5.6) Apply the solution process of this chapter to find an approximate solution of the problem

$$\mathcal{L}u \equiv u_{xx} - u_t = 0$$

$$u(0, t) = e^{-t}, \qquad u_x(1, t) = 0$$

$$u(x, 0) = 1.$$

5.7) Apply the solution process of this chapter to find an approximate solution of the problem

$$\mathcal{L}u \equiv u_{xx} - u_t = 2$$

$$u(0, t) = u(1, t)$$

$$u_x(0, t) = u_x(1, t)$$

$$u(x, 0) = \sin 2\pi x.$$

5.8) Apply the solution process of this chapter to find an approximate solution of the problem

$$\mathcal{L}u \equiv u_{xx} - u_t = xt$$

$$u_x(0,t) = 2, \qquad u_x(1,t) = 1$$
$$u(x,0) = 2x - x^2/2.$$

5.9) Determine $F(x,t)$, $u_0(x)$, $A(t)$, and $B(t)$ such that

$$u(x,t) = xe^{-t}$$

is a solution of

$$\mathcal{L}u = u_{xx} - u_t = F(x,t)$$
$$u(0,t) = A(t), \qquad u(1,t) = B(t)$$
$$u(x,0) = u_0(x).$$

Use the solution process of this chapter to find an approximate solution u_N. Compute $e(x,t) = P_N u(x,t) - u_N(x,t)$.

5.10) Determine $F(x,t)$, $u_0(x)$, $A(t)$, and $B(t)$ such that

$$u(x,t) = (x-t)^2$$

solves

$$\mathcal{L}u \equiv u_{xx} - u_t = F(x,t)$$
$$u(0,t) = A(t), \qquad u_x(1,t) = B(t)$$
$$u(x,0) = u_0(x).$$

Use the solution process of this chapter to find an approximate eigenfunction solution u_N. Compute $e(x,t) = P_N u(x,t) - u_N(x,t)$.

5.11) Apply the solution process of this chapter to find an approximate solution of the problem

$$\mathcal{L}u \equiv u_{xx} - u_t - tu = 0$$
$$u(0,t) = 1, \qquad u(1,t) = 0$$
$$u(x,0) = 1 - x.$$

5.12) Let C be continuously differentiable with respect to t for $t > 0$. Show that the initial value problem

$$-\lambda_n^2 \alpha_n - \alpha_n' = C'(t)\hat{\gamma}_n$$

$$\alpha_n(0) = \hat{\alpha}_n$$

can be solved in terms of $C(t)$ without ever computing $C'(t)$.

5.13) Apply the solution process of this chapter to find an approximate solution of the problem

$$\mathcal{L}u \equiv u_{xx} - u_{tt} = 0$$

$$u(0,t) = 0, \qquad u_x(1,t) = 0$$

$$u(x,0) = \sin\left(\frac{\pi}{2}x\right)$$

$$u_t(x,0) = 0.$$

5.14) Apply the solution process of this chapter to find an approximate solution of the problem

$$\mathcal{L}u \equiv u_{xx} - u_{tt} = tx(1 - x)$$

$$u(0,t) = u(1,t) = 0$$

$$u(x,0) = 0$$

$$u_t(x,0) = \sin \pi x.$$

5.15) Determine $F(x,t)$, $u_0(x)$, $u_1(x)$, $A(t)$, and $B(t)$ such that

$$u(x,t) = (1 - x)\sin \omega t$$

solves

$$\mathcal{L}u \equiv u_{xx} - u_{tt} = F(x,t)$$

$$u(0,t) = A(t), \qquad u(1,t) = B(t)$$

$$u(x,0) = u_0(x)$$

$$u_t(x,0) = u_1(x).$$

Then use the solution process of this chapter to find an approximate solution u_N of this problem and compare it with the analytic solution of the original problem.

5.16) Let C be twice continuously differentiable with respect to t for $t > 0$. Show that the initial value problem

$$-\lambda_n^2 c^2 \alpha_n - \alpha_n'' = C''(t)\hat{\gamma}_n$$

$$\alpha_n(0) = \hat{a}_n, \qquad \alpha_n'(0) = \hat{\beta}_n$$

can be solved in terms of $C(t)$ without ever computing $C''(t)$.

5.17) Apply the solution process of this chapter to find an approximate solution of the vibrating beam problem

$$\mathcal{L}u \equiv u_{xxxx} + u_{tt} = 0$$

$$u(0, t) = \sin t, \qquad u(1, t) = 0$$

$$u_{xx}(0, t) = u_{xx}(1, t) = 0$$

$$u(x, 0) = 0$$

$$u_t(x, 0) = (1 - x)$$

(see Exercise 3.14).

5.18) Apply the solution process of this chapter to find an approximate solution of the problem

$$\mathcal{L}u \equiv u_{xx} + u_{yy} = 1$$

$$u(0, y) = y, \qquad u(2, y) = 1 - y$$

$$u_y(x, 0) = 1 - 2x, \qquad u_y(x, 1) = \cos \pi x.$$

Solve this problem twice: once with eigenfunctions of the independent variable x, the second time with eigenfunctions of the independent variable y.

5.19) Apply the solution process of this chapter to find an approximate solution of the problem

$$\mathcal{L}u \equiv u_{xx} + u_{yy} + u_y = 1$$

$$u(0, y) = u(1, y) = 0$$

$$u(x, 0) = u(x, 1) = 0.$$

5.20) Determine $F(x, y)$ and $g(x, y)$ such that

$$u(x, y) = (x + y)^2$$

is a solution of

$$\Delta u = F(x, y), \qquad (x, y) \subset D = (0, 2) \times (0, 1)$$

$$u = g, \qquad (x, y) \in \partial D.$$

Now solve this Dirichlet problem in the following two ways:

i) Use the solution process of this chapter to find an approximate solution $u_N(x, y)$ in terms of eigenfunctions which are functions of x. Compute $e(x, y) = P_N u(x, y) - u_N(x, y)$.

ii) Use the solution process of this chapter to find an approximate solution $u_N(x, y)$ in terms of eigenfunctions which are functions of y. Compute $e(x, y) = P_N u(x, y) - u_N(x, y)$.

5.21) Determine $F(x, y)$ and $g(x, y)$ such that

$$u(x, y) = (xy)^2$$

is a solution of

$$\Delta u = F(x, y), \qquad (x, y) \subset D = (0, 2) \times (0, 1)$$

$$\frac{\partial u}{\partial n} = g, \qquad (x, y) \in \partial D.$$

Now solve this Neumann problem in the following two ways:

i) Use the solution process of this chapter to find an approximate solution $u_N(x, y)$ in terms of eigenfunctions which are functions of x. Compute $e(x, y) = P_N u(x, y) - u_N(x, y)$.

ii) Use the solution process of this chapter to find an approximate solution $u_N(x, y)$ in terms of eigenfunctions which are functions of y. Compute $e(x, y) = P_N u(x, y) - u_N(x, y)$.

Chapter 6

One-Dimensional Diffusion Equation

The general solution process of the last chapter will be applied to diffusion problems of increasing complexity. Our aim is to demonstrate that separation of variables, in general, and the eigenfunction expansion method, in particular, can provide quantitative and numerical answers for a variety of realistic problems. These problems are usually drawn from conductive heat transfer, but they have natural analogues in other diffusion contexts, such as mass transfer, flow in a porous medium, and even option pricing. The chapter concludes with some theoretical results on the convergence of the approximate solution to the exact solution and on the relationship between the eigenfunction expansion and Duhamel's superposition method for problems with time-dependent source terms.

6.1 Applications of the eigenfunction expansion method

Example 6.1 How many terms of the series solution are enough?

At the end of this chapter we shall discuss some theoretical error bounds for the approximate solution of the heat equation. However, for some problems very specific information is available which can provide insight into the solution process and the quality of the answer. To illustrate this point we shall consider the model problem

$$\mathcal{L}u \equiv u_{xx} - \frac{1}{\alpha} u_t = 0, \qquad 0 < x < L, \quad t > 0$$

$$u(0,t) = 1, \quad u(L,t) = 0, \quad t > 0$$

$$u(x,0) = 0, \quad\quad 0 < x < L.$$

This is a common problem in every text on separation of variables. It describes a thermal system initially in a uniform state which is shocked at time $t = 0$ with an instantaneous temperature rise. The parameter α in the above heat equation is the so-called diffusivity of the medium and is included (rather than set to 1 by rescaling time) to show explicitly the dependence of u on α.

It is known from the theory of partial differential equations that this problem has a unique infinitely differentiable solution $u(x,t)$ on $(0,L) \times (0,T]$ for all $T > 0$, and which takes on the boundary and initial conditions. However, since

$$\lim_{t \to 0} u(0,t) \neq \lim_{x \to 0} u(x,0)$$

the solution is discontinuous at $(0,0)$. As we shall see this discontinuity will introduce a Gibbs phenomenon into our approximating problem. We shall compute an approximate solution $u_N(x,t)$ and would like to get an idea of how large N should be in order to obtain a good solution.

The problem is transformed to one with zero boundary data by subtracting the steady-state solution

$$v(x) = \left(1 - \frac{x}{L}\right).$$

We set

$$w(x,t) = u(x,t) - v(x).$$

Then

$$\mathcal{L}w = \mathcal{L}u - \mathcal{L}v = 0$$

$$w(0,t) = w(L,t) = 0$$

$$w(x,0) = -v(x).$$

The associated eigenvalue problem is

$$\phi'' = \mu\phi$$

$$\phi(0) = \phi(L) = 0.$$

The eigenfunctions are $\phi_n(x) = \sin\lambda_n x$ with $\lambda_n = \frac{n\pi}{L}$ and $\mu_n = -\lambda_n^2$. The approximating problem in the span of the first N eigenfunctions is readily found. In this case the source term G is zero so that

$$P_N G(x,t) = \sum_{n=1}^{N} \gamma_n(t)\phi_n(x) \equiv 0,$$

i.e.

$$\gamma_n(t) \equiv 0 \quad \text{for all } n.$$

The projection of the initial condition is

$$P_N w(x, 0) = P_N(-v(x)) = \sum_{n=1}^{N} \hat{\alpha}_n \phi_n(x)$$

with

$$\hat{\alpha}_n = \frac{\langle -v(x), \phi_n \rangle}{\langle \phi_n, \phi_n \rangle} = -\frac{2}{L} \int_0^L \left(1 - \frac{x}{L}\right) \sin \lambda_n x \, dx = -\frac{2}{L \lambda_n}.$$

Since the odd extension of $-v(x)$ to $[-L, L]$ has a jump at $x = 0$, we expect a Gibbs phenomenon in the approximation of $v(x)$ in terms of the $\{\phi_n\}$.

The solution of

$$\mathcal{L}w = 0$$

$$w(0, t) = w(L, t) = 0$$

$$w(x, 0) = -P_N(v(x))$$

is given by

$$w_N(x, t) = \sum_{n=1}^{N} \alpha_n(t) \phi_n(x)$$

where

$$-\lambda_n^2 \alpha_n - \frac{1}{\alpha} \alpha_n' = 0.$$

It follows from (5.8) that the exact solution of the approximating problem is

$$u_N(x, t) = -\sum_{n=1}^{N} \frac{2}{n\pi} e^{-\lambda_n^2 \alpha t} \sin \lambda_n x + \left(1 - \frac{x}{L}\right). \tag{6.1}$$

The time-dependent terms in (6.1) constitute the so-called transient part of the solution. The infinite series obtained from (6.1) as $N \to \infty$ is generally considered the separation of variables solution of the problem, but only the finite sum in (6.1) can be computed. In practice $u_N(x, t)$ is evaluated for a given x and t and a few N. If changes in the answer with N become insignificant, the last computed value is accepted as the solution of the original problem. However, for small t this N can be quite large as the following argument shows.

We know from our discussion of the Gibbs phenomenon that $u_N(x, 0)$ converges to $u_0(x)$ only pointwise on $(0, L]$ as $N \to \infty$ and that

$$\max_x [u_N(x, 0) - u_0(x)] \cong .179 \quad \text{as } N \to \infty.$$

We shall now show that for $t > 0$ the approximate solution $u_N(x, t)$ converges uniformly on $[0, L]$ as $N \to \infty$. Let $N > M$, then

$$|u_N(x, t) - u_M(x, t)| = \left| \sum_{n=M+1}^{N} \frac{2}{n\pi} e^{-(\frac{\pi}{L})^2 \alpha t n^2} \phi_n(x) \right|$$

$$\leq \sum_{n=M+1}^{\infty} \frac{2}{(M+1)\pi} A(\alpha t, M+1)^n$$

$$= \frac{2}{(M+1)\pi} A(\alpha t, M+1)^{M+1} \frac{1}{1 - A(\alpha t, M+1)}$$

$$\equiv R(\alpha t, M+1)$$

where

$$A(\alpha t, M) = e^{-(\frac{\pi}{L})^2 \alpha t M} < 1 \qquad \text{for all } t > 0.$$

We see that $R \to 0$ as $M \to \infty$ for all $N > M$ independently of $x \in [0, L]$. If $\lim_{N \to \infty} u_N(x, t)$ is accepted as the analytic solution of the original problem, then $R(\alpha t, M+1)$ is a bound on the error. We can estimate M such that $R < \epsilon$ for any given $\epsilon > 0$. For example, suppose we wish to assure that our solution at time $\alpha t = .00001$ is within 10^{-6} of the analytic solution. With $L = 1$ we find that

$$R(10^{-5}, 334) < 10^{-6} < R(10^{-5}, 333).$$

Hence 333 terms in the transient solution are sufficient for the approximate solution. Of course, our estimates are not sharp, but we are not far off the mark. For example, a numerical evaluation of (6.1) for $N = 300$ and $\alpha t = 10^{-5}$ yields

$$\min_x u_{300}(x, t) = -2.74 \times 10^{-6}.$$

Since the analytic solution is nonnegative, the error exceeds our tolerance. (We remark that for $N = 333$

$$\min u_{333}(x, t) = -0.83 \times 10^{-6}$$

is within our tolerance.)

For illustration we show in Fig. 6.1 $u_{10}(x, t)$ and $u_{333}(x, t)$ for $\alpha t = 10^{-5}$ to caution that one cannot always truncate the series after just a few terms.

In contrast, similar estimates for $\alpha t = .1$ show that only three terms are required in the transient solution for an error less than 10^{-6}.

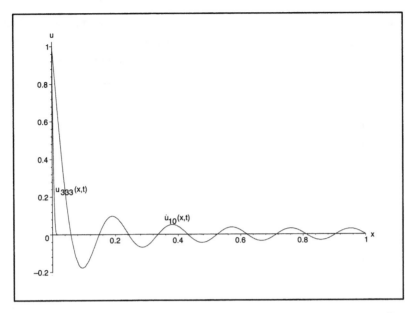

Figure 6.1: [1]Plot of $u_N(x, t)$ for $N = 10$ and $N = 333$ at $\alpha t = 10^{-5}$.

Example 6.2 Determination of an unknown diffusivity from measured data.

The advantage of an analytic solution is particularly pronounced when it comes to an estimation of parameters in the equation from observed data. We shall illustrate this point with the model problem of Example 6.1, but this time we shall assume that the diffusivity of the medium is not known. Instead at time $t = 1$ we have temperature measurements

$$m(1/4) = .4$$

$$m(1/2) = .1$$

$$m(3/4) = .01$$

recorded at $x = 1/4$, $x = 1/2$, and $x = 3/4$. We want to find a constant diffusivity α for which (6.1) best matches the measurements in the least squares sense, i.e., we need to minimize

$$E(\alpha) = \sum_{i=1}^{3} (u(x_i, 1, \alpha) - m(x_i))^2$$

---

[1]Subsequently, figures and tables are numbered according to the examples in which they appear.

where $x_i = i/4$, $i = 1, 2, 3$. Considering the crudeness of the model we shall be content to graph $E(\alpha)$ vs. $y = e^\alpha$ and read off where it has a minimum. Fig. 6.2 shows $E(\alpha)$ when $N = 5$. There appears to be a unique minimum.

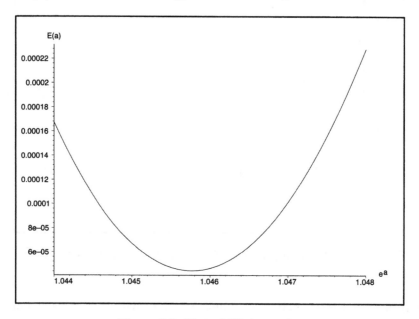

Figure 6.2: Plot of $E(\alpha)$ vs. e^α.

The diffusivity minimizing the error is observed to be

$$\alpha = .0448.$$

The temperatures predicted by this α are

$$u(1/4, 1) = .403, \quad u(1/2, 1) = .095, \quad u(3/4, 1) = .0122.$$

$t = 1$ is large enough that increasing the number of terms in our approximate solution does not change the answer. In fact, $N = 5$ is consistent with an error of less than 10^{-6} as discussed in Example 6.1.

Example 6.3 Thermal waves.

Our next example introduces flux data at $x = L$.
We consider

$$\mathcal{L}u \equiv u_{xx} - u_t = 0$$

$$u(0, t) = 1, \quad u_x(L, t) = 0$$

$$u(x, 0) = 0.$$

This is a standard companion problem to that of Example 6.1. It describes heat flow in a slab or axial flow in a bar where the right end of the slab or bar is perfectly insulated. Its solution is straightforward. The boundary data are zeroed out by choosing

$$v(x) = 1$$

(again the steady-state solution of the problem) and setting

$$w(x,t) = u(x,t) - v(x).$$

Then

$$\mathcal{L}w = 0$$

$$w(0,t) = w_x(L,t) = 0$$

$$w(x,0) = -1.$$

The associated eigenvalue problem is

$$\phi'' = \mu\phi$$

$$\phi(0) = \phi'(L) = 0$$

which has the solutions

$$\phi_n(x) = \sin\lambda_n x, \quad \lambda_n = \frac{\left(\frac{\pi}{2} + n\pi\right)}{L}, \quad \mu_n = -\lambda_n^2, \quad n = 0, 1, \ldots.$$

The approximating problem is

$$\mathcal{L}w = 0$$

$$w(0,t) = w_x(L,t) = 0$$

$$w(x,0) = P_N(-1) = \sum_{n=0}^{N} \hat{a}_n \phi_n(x)$$

where

$$\hat{a}_n = \frac{\langle -1, \phi_n \rangle}{\langle \phi_n, \phi_n \rangle} = -\frac{2}{L}\int_0^L \sin\lambda_n x \, dx = \frac{-2}{\lambda_n L}.$$

We expect a Gibbs phenomenon in the approximation to $w(x,0)$ at $x = 0$. The approximate solution to our problem is

$$u_N(x,t) = \sum_{n=0}^{N} \hat{a}_n e^{-\lambda_n^2 t}\phi_n(x) + 1. \tag{6.2}$$

The simple formula (6.2) has a surprising consequence. The solution $u_N(x,t)$ at $x = L$ is seen to be a linear combination of the functions

$$\left\{1, e^{-\lambda_0^2 t}, \ldots, e^{-\lambda_N^2 t}\right\}.$$

But these functions are linearly independent. (Their Wronskian at $t = 0$ is the nonzero determinant of a Vandermonde matrix.) Hence there is no interval $[0, T]$ for $T > 0$ such that

$$u_N(L, t) \equiv 0 \quad \text{for } t \in [0, T].$$

In other words, the thermal signal generated by the boundary condition

$$u(t, 0) = 1, \qquad t > 0,$$

is felt immediately at $L = 1$. Thus the thermal signal travels with infinite speed through $[0, L]$. This phenomenon was already observed in Section 1.4 and is a well-known consequence of Fourier's law of heat conduction. It contradicts our experience that it will take time before the heat input at $x = 0$ will be felt at $x = L$. But in fact, for all practical purposes a detectable signal does travel with finite speed. Fig. 6.3a shows a plot of $u_N(x, t)$ for a few values of t.

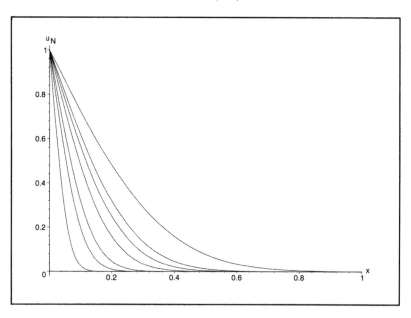

Figure 6.3: (a) Plot of $u_N(x, t)$ for increasing t.

We see a distinct thermal wave moving through the interval. We can use (6.2) to compute the speed with which an isotherm moves through the interval. We shall set $L = 1$ and determine the speed $s'(t)$ of the isotherm

$$u_N(s(t), t) = .001.$$

Specifically, for given $x_i = i/100$, $i = 10, \dots, 99$ we shall compute first the time $t = T(x_i)$ when the isotherm reaches x_i. $T(x_i)$ is found by applying Newton's method to the nonlinear equation

$$F(y) \equiv u_N(x_i, t) - .001 = 0$$

where $y = e^t$ and $u_N(x, t)$ is given by (6.2) so that

$$F(y) = \sum_{n=0}^{N} \hat{a}_n y^{-\lambda_n^2} \phi_n(x_i) + .999.$$

Once $T(x_i)$ is known the speed $s'(T(x_i))$ of the isotherm at x_i can be determined from

$$\frac{d}{dt} u_N(s(t), t) = \frac{\partial}{\partial x} u_N(s(t), t) s'(t) + \frac{\partial}{\partial t} u_N(s(t), t) = 0$$

for $s(t) = x_i$ and $t = T(x_i)$. The partial derivatives of u_N are available from (6.2).

Figs. 6.3b,c show $T(x_i)$ and $s'(t)$ for points in the interval $(0, 1)$.

The answers remain unchanged for $N > 50$. However, we need to point out that we begin our calculation at $x = .1$. For $x < .1$ the initial Gibbs phenomenon causes very slow convergence of the series (6.2) because $t = T(x)$ is small. This makes the calculation of $T(x)$ difficult and the answers unreliable. However, because the answer is analytic, this initial effect does not pollute the answer at later times and other locations.

To give meaning to these numbers suppose a 1 m copper rod is insulated along its length and at its right end. Its initial temperature is $0°C$. For time $\tau > 0$ its temperature at the left end is maintained at $100°C$. The thermal model for the temperature $T(y, \tau)$ in the rod is

$$T_{yy}(y, \tau) - \frac{1}{\alpha} T_\tau(y, \tau) = 0$$

$$T(0, \tau) = 100, \qquad T_y(100, \tau) = 0$$

$$T(y, 0) = 0$$

where α is the diffusivity of copper, given in [19] as

$$\alpha \cong 1.1 \text{ cm}^2/\text{sec}.$$

If we write $u(x, t) = T(y(x), \tau(t))/100$, $x = y/100$, $t = \alpha 10^{-4}\tau$, then u satisfies the equations of this example. The isotherm $u = .001$ (i.e., $T = .1°C$) reaches the right end at $t \cong 0.0412$ which corresponds to a real time of

$$\tau \cong 375 \text{ sec}.$$

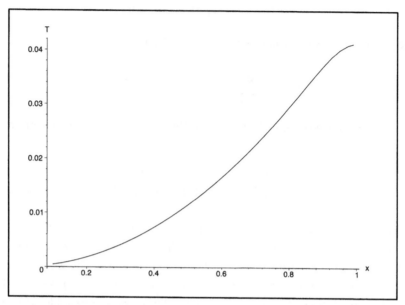

Figure 6.3: (b) Arrival time T of the isotherm $u = .001$ at x.

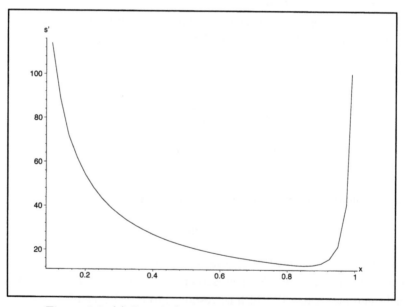

Figure 6.3: (c) Speed s' of the isotherm $u = .001$ at x.

Conversely, the transient part of the solution will have decayed at the end of the rod to $-.001$ when

$$-.001 \cong \alpha_0 e^{-\lambda_0^2 t} \phi_0(1)$$

which yields $t = 2.89$ and a real time of $\tau = 7.3$ hours. The additional terms of the transient solution are negligible for such large t.

Example 6.4 Matching a temperature history.

We shall determine a heating schedule at $x = 0$ to match a desired temperature at the end of a perfectly insulated slab or bar. Specifically, we shall find a polynomial $P_K(t)$ with $P_K(0) = 0$ such that the solution $u(x,t)$ of

$$\mathcal{L}u \equiv u_{xx} - u_t = 0$$

$$u(0,t) = P_K(t), \quad u_x(1,t) = 0$$

$$u(x,0) = 0$$

minimizes the integral

$$\int_0^T (g(t) - u(L,t))^2 dt$$

where g is a given target function and T is fixed.

In general, this is a difficult problem and may not have a computable solution. However, the corresponding approximating problem is straightforward to solve.

The boundary conditions are zeroed out if we choose

$$v(x,t) = v(t) = P_K(t) = \sum_{j=1}^{K} c_j t^j$$

where the $\{c_j\}$ are to be determined. Then

$$w(x,t) = u(x,t) - v(t)$$

satisfies

$$\mathcal{L}w = \mathcal{L}u - \mathcal{L}v = v_t = P_K'(t)$$

$$w(0,t) = w_x(1,t) = 0$$

$$w(x,0) = 0.$$

The associated eigenvalue problem is the same as in Example 6.2, and the approximating problem is

$$\mathcal{L}w = \sum_{n=0}^{N} \hat{\gamma}_n P_K'(t) \phi_n(x)$$

$$w(x, 0) = 0$$

where

$$\hat{\gamma}_n = \frac{2}{L} \langle 1, \phi_n \rangle = \frac{2}{L\lambda_n}.$$

The problem is solved by

$$w_N(x, t) = \sum_{n=0}^{N} \alpha_n(t)\phi_n(x)$$

where

$$-\lambda_n^2 \alpha_n - \alpha_n' = \hat{\gamma}_n P_K'(t)$$

$$\alpha_n(0) = 0.$$

The variation of parameters solution is

$$\alpha_n(t) = -\int_0^t \hat{\gamma}_n e^{-\lambda_n^2(t-s)} P_K'(s) ds$$

so that

$$w_N(x, t) = -\sum_{j=1}^{K} \sum_{n=0}^{N} c_j \left(\int_0^t \hat{\gamma}_n e^{-\lambda_n^2(t-s)} j s^{j-1} ds \right) \phi_n(x).$$

Hence the approximate solution

$$u_N(x, t) = w_N(x, t) + P_K(t)$$

at $x = 1$ is

$$u_N(1, t) = \sum_{j=1}^{K} D_j(t) c_j$$

where

$$D_j(t) = -j \sum_{n=0}^{N} \left[\int_0^t \hat{\gamma}_n e^{-\lambda_n^2(t-s)} s^{j-1} ds \phi_n(1) \right] + t^j.$$

It is now straightforward to minimize

$$E(c_1, \ldots, c_K) = \int_0^T \left(g(t) - \sum_{j=1}^{K} D_j(t) c_j \right)^2 dt.$$

Calculus shows that the solution $\vec{c}$ is found from the matrix system

$$\mathcal{A}\vec{c} = \vec{b}$$

where
$$A_{ij} = \int_0^T D_i(t)D_j(t)dt$$
and
$$b_i = \int_0^T D_i(t)g(t)dt.$$
For a representative calculation we choose $L = T = 1$ and try to match

$$g(t) = \sin \pi t.$$

Fig. 6.4a shows the optimal polynomial $P_4(t)$ while Fig. 6.4b shows the target function $g(t)$ and the computed approximation $u_N(x,t)$ for $N = 20$.

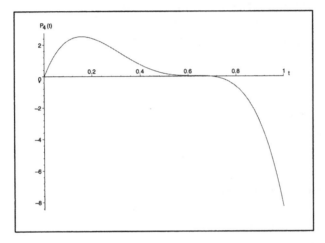

Figure 6.4: (a) Computed polynomial $P_4(t)$ for the boundary condition $u(0,t) = P(t)$.

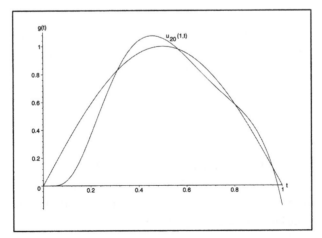

Figure 6.4: (b) Plot of $u_{20}(1,t)$ and the target function $g(t) = \sin \pi t$.

The computation was carried out with Maple which automatically performs all calculations symbolically where possible and numerically otherwise. In this language the entire problem took no more programming effort than writing the above mathematical expressions in Maple notation. To illustrate this point we list below the entire Maple program leading to the graphs in Figure 6.4a,b.

```
> with(LinearAlgebra):with(plots):
> lambda:=n->(2*n+1)*(Pi/2);
> phi:=(x,n)->sin(lambda(n)*x);
> g:=n->integrate(phi(x,n),x=0..1)/integrate(phi(x,n)^2,x=0..1);
> F:=(t,N,k)->k*add(-g(n)*integrate(exp((s-t)*lambda(n)^2)*
      (s^(k-1)),s=0..t)* phi(1,n),n=0..N)+t^k;
> a:=(j,k)->integrate(F(t,N,k)*F(t,N,j),t=0..T);
> b:=j->integrate(B(t)*F(t,N,j),t=0..T);
> N:=10;K:=4;T:=1;
> B:=t->sin(Pi*t);
> AA:=evalf(Matrix(K,K,(j,k)->a(j,k)));
> RS:=evalf(Matrix(K,1,(j)->b(j)));
> COE:=LinearSolve(AA,RS);
> P:=t->add(COE[j,1]*t^j,j=1..K);
> plot(P(t),t=0..1);
> PP:=t->add(j*COE[j,1]*t^(j-1),j=1..K);
> alpha:=(t,n)->-g(n)*integrate(PP(s)*exp(-(t-s)*
  lambda(n)^2),s=0..t);
> u:=t->add(alpha(t,n)*phi(1,n),n=0..20)+P(t);
> p1:=plot(u(t),t=0..1):p2:=plot(B(t),t=0..1):
> display(p1,p2);
```

The above calculation yields a polynomial of degree K whose coefficients c_j clearly depend on the dimension of the subspace in which the solution $w_N(x,t)$ is found, i.e., $c_j = c_j(N)$. In order for the answer to remain meaningful for the original problem, we would need a proof that

$$\lim_{N\to\infty} c_j(N) = c_j^*, \qquad j = 1,\ldots,K$$

for constants $\{c_j^*\}$. We do not have such a result but take assurance from numerical experiments that show that $P_4(t)$ changes little for $N > 10$.

Such a behavior cannot always be expected, and it may well be that a problem is solvable in the subspace but that the answers diverge as $N \to \infty$. For example, consider the following problem:

$$\mathcal{L}u \equiv u_{xx} - u_t = 0$$

$$u(0,t) = u_x(1,t) = 0$$

$$u(x,0) = u_0(x)$$

where u_0 is to be determined such that

$$\|u(x,T) - u_F(x)\|^2 \equiv \int_0^1 (u(x,T) - u_F(x))^2 dx$$

is minimized for a given target function u_F. This formulation is actually a disguised way of writing an initial value problem for the backward heat equation which, we know from Section 1.4, is a notorious ill-posed problem. If $u_F(x)$ is projected into the subspace spanned by the first N eigenfunctions of the associated Sturm–Liouville problem, i.e., if we wish to find $u_0(x)$ such that

$$\|u(x,T) - P_N u_F(x)\|$$

is minimized where

$$P_N u_F(x) = \sum_{n=1}^{N} \hat{\beta}_n \phi_n(x),$$

then

$$u_N(x,T) = \sum_{n=1}^{N} \alpha_n(T)\phi_n(x) \equiv P_N u_F(x)$$

if

$$\alpha_n(0) = \hat{\beta}_n e^{\lambda_n^2 T},$$

i.e., we have

$$\|u_N(x,T) - P_N u_F(x)\| = 0.$$

But u_N does not converge for $t < T$ as $N \to \infty$ because $\alpha_n(t) \to \infty$ as $n \to \infty$.

Example 6.5 Phase shift for a thermal wave.

Consider the problem

$$\mathcal{L}u \equiv u_{xx} - u_t = 0$$

$$u(0,t) = \sin \omega t, \quad u_x(1,t) = 0$$

$$u(x,0) = 0.$$

It is reasonable to expect that $u(1,t)$ will vary sinusoidally with frequency ω as $t \to \infty$. Our aim is to find the phase shift of $u(1,t)$ relative to $u(0,t)$.

To zero out the boundary data we set $w(x,t) = u(x,t) - \sin \omega t$. Then

$$\mathcal{L}w \equiv w_{xx} - w_t = \omega \cos \omega t$$

$$w(0, t) = w_x(1, t) = 0$$

$$w(x, 0) = 0.$$

As above the eigenfunctions and eigenvalues are

$$\phi_n(x) = \sin \lambda_n x \quad \text{where } \lambda_n = \left(\frac{\pi}{2} + n\pi \right), \quad \mu_n = -\lambda_n^2, \quad n = 0, 1, \dots.$$

Then

$$P_N(\omega \cos \omega t) = \omega \cos \omega t \sum_{n=0}^{N} \hat{\gamma}_n \phi_n(x)$$

where

$$\hat{\gamma}_n = \frac{\langle 1, \phi_n \rangle}{\langle \phi_n, \phi_n \rangle} = \frac{2}{\lambda_n}.$$

If

$$w_N(x, t) = \sum_{n=0}^{N} \alpha_n(t) \phi_n(x),$$

then

$$-\lambda_n^2 \alpha_n(t) - \alpha_n'(t) = \hat{\gamma}_n \omega \cos \omega t.$$

It follows that

$$\alpha_n(t) = c_n e^{-\lambda_n^2 t} + \alpha_{np}(t).$$

To find a particular integral we use the method of undetermined coefficients and try

$$\alpha_{np}(t) = A_n \cos \omega t + B_n \sin \omega t.$$

Substituting into the differential equation and equating the coefficients of $\sin \omega t$ and $\cos \omega t$ we find

$$-\lambda_n^2 A_n - \omega B_n = \hat{\gamma}_n \omega$$

$$-\lambda_n^2 B_n + \omega A_n = 0$$

so that

$$A_n = \frac{-\hat{\gamma}_n \omega \lambda_n^2}{\lambda_n^4 + \omega^2}$$

$$B_n = \frac{-\gamma_n \omega^2}{\lambda_n^4 + \omega^2}.$$

The final approximate answer to our problem is

$$u_N(x, t) = \sum_{n=0}^{N} \left[A_n \left(\cos \omega t - e^{-\lambda_n^2 t} \right) + B_n \sin \omega t \right] \phi_n(x) + \sin \omega t.$$

The exponential terms decay rapidly and will be ignored. It is now straightforward to express the phase shift in terms of

$$C_N = \sum_{n=0}^{N} A_n \phi_n(1) \quad \text{and} \quad D_N = \sum_{n=0}^{N} B_n \phi_n(1) + 1.$$

However, perhaps more revealing is the following approach. We write

$$\sin \omega t \sim \sin \omega t \sum_{n=0}^{N} \hat{\gamma}_n \phi_n(x).$$

This is the Fourier series of the four-periodic odd function which coincides with $\sin \omega t$ on $(0, 2)$. This series converges uniformly near $x = 1$. Ignoring again the exponential terms we can write

$$u_N(1, t) \sim \sum_{n=0}^{N} [A_n \cos \omega t + (B_n + \hat{\gamma}_n) \sin \omega t] \phi_n(1).$$

This expression can be rearranged into

$$u_N(1, t) \sim \sum_{n=0}^{N} \frac{\hat{\gamma}_n \lambda_n^2}{\sqrt{\lambda_n^4 + \omega^2}} \left[-\frac{\omega}{\sqrt{\lambda_n^4 + \omega^2}} \cos \omega t + \frac{\lambda_n^2}{\sqrt{\lambda_n^4 + \omega^2}} \sin \omega t \right] \phi_n(1).$$

If we set

$$\sin \psi_n = \frac{\omega}{\sqrt{\lambda_n^4 + \omega^2}},$$

then

$$u_N(1, t) = \sum_{n=0}^{N} -\frac{\hat{\gamma}_n \lambda_n^2}{\sqrt{\lambda_n^4 + \omega^2}} \sin(\omega t - \psi_n) \phi_n(1).$$

Since $\hat{\gamma}_n = \frac{2}{\lambda_n}$, we see that the dominant term corresponds to $n = 0$ which yields a phase shift ψ_0 given by

$$\sin \psi_0 = \frac{\omega}{\sqrt{\left(\frac{\pi}{2}\right)^4 + \omega^2}}.$$

Example 6.6 Dynamic determination of a convective heat transfer coefficient from measured data.

A bar insulated along its length is initially at the uniform ambient temperature T_∞ and then heated instantaneously at one end to $T_0 > T_\infty$ while it loses energy at the other end due to convective cooling. The aim is to find a heat transfer coefficient consistent with measured temperature data at that end. We may assume that after scaling space and time the nondimensional temperature

$$u(x,t) = \frac{T(x,t) - T_\infty}{T_0 - T_\infty}$$

satisfies the problem

$$\mathcal{L}u \equiv u_{xx} - u_t = 0$$

$$u(0,t) = 1, \qquad u_x(1,t) = -hu(1,t)$$

$$u(x,0) = 0$$

where h is an unknown (scaled) heat transfer coefficient which is to be determined such that $u(1,t)$ is consistent with measured data $(t_i, U(t_i))$, where $U(t_i)$ is the temperature recorded at $x = 1$ and $t = t_i$ for $i = 1, \ldots, M$.

It is reasonable to suggest that h should be computed such that $u(1, t_i)$ approximates $U(t_i)$ in the mean square sense. Hence we wish to find that value of h which minimizes

$$E(h) = \sum_{i=1}^{M} (u(1, t_i, h) - U(t_i))^2 \hat{w}(t_i)$$

where $u(x,t,h)$ indicates that the analytic solution u depends on h. $\hat{w}(t)$ is a weight function chosen to accentuate those data which are thought to be most relevant. The relationship between $u(x,t,h)$ and h is quite implicit and nonlinear so that the tools of calculus for minimizing $E(h)$ are of little use. However, it is easy to calculate and plot $E(h)$ for a range of values for h if we approximate u by its eigenfunction expansion. To find $u_N(x,t)$ we write

$$w(x,t) = u(x,t,h) - v(x)$$

where

$$v(x) = 1 - \frac{h}{1+h}x.$$

Then

$$\mathcal{L}w \equiv w_{xx} - w_t = 0$$

$$w(0,t) = 0, \qquad w_x(1,t) = -hw(1,t)$$

$$w(x,0) = \frac{h}{1+h}x - 1.$$

The associated eigenvalue problem is

$$\phi'' = \mu\phi$$

$$\phi(0) = 0, \quad \phi'(1) = -h\phi(1).$$

The eigenfunctions are

$$\phi_n(x) = \sin \lambda_n x$$

where $\{\lambda_n(h)\}$ are the solutions of

$$f(\lambda, h) = \lambda \cos \lambda + h \sin \lambda = 0. \qquad (6.3)$$

For $h = 0$ the roots are $\lambda_n(0) = \frac{\pi}{2} + (n-1)\pi$ for $n = 1, 2, \ldots$. Newton's method will yield the corresponding $(\lambda_n(h_k))$ for $h_k = h_{k-1} + \Delta h$ with Δh sufficiently small when $\lambda_n(h_{k-1})$ is chosen as an initial guess for the iteration. Table 6.6 below contains some representative results.

Table 6.6: Roots of $f(\lambda, h) = 0$

h	λ_1	λ_2	λ_3	λ_4	λ_5
0.00000	1.57080	4.71239	7.85398	10.99557	14.13717
0.10000	1.63199	4.73351	7.86669	11.00466	14.14424
0.20000	1.68868	4.75443	7.87936	11.01373	14.15130
0.30000	1.74140	4.77513	7.89198	11.02278	14.15835
0.40000	1.79058	4.79561	7.90454	11.03182	14.16540
0.50000	1.83660	4.81584	7.91705	11.04083	14.17243
0.60000	1.87976	4.83583	7.92950	11.04982	14.17946
0.70000	1.92035	4.85557	7.94189	11.05879	14.18647
0.80000	1.95857	4.87504	7.95422	11.06773	14.19347
0.90000	1.99465	4.89425	7.96648	11.07665	14.20046
1.00000	2.02876	4.91318	7.97867	11.08554	14.20744

The eigenfunction expansion for w is

$$w_N(x, t) = \sum_{n=1}^{N} \alpha_n(t) \phi_n(x)$$

where

$$-\lambda_n^2 \alpha_n(t) - \alpha_n'(t) = 0$$

$$\alpha_n(0) = \frac{h}{1+h} \frac{\langle x, \phi_n \rangle}{\langle \phi_n, \phi_n \rangle} - \frac{\langle 1, \phi_n \rangle}{\langle \phi_n, \phi_n \rangle}.$$

A straightforward integration and the use of (6.3) show that

$$\frac{h}{1+h} \langle x, \phi_n \rangle = -\frac{1}{\lambda_n} \cos \lambda_n$$

$$\langle 1, \phi_n \rangle = \frac{1}{\lambda_n} (1 - \cos \lambda_n)$$

$$\langle \phi_n, \phi_n \rangle = \frac{1}{2} \left(1 + \frac{1}{h} \cos^2 \lambda_n \right)$$

so that

$$\alpha_n(t) = \frac{-2h}{\lambda_n (h + \cos \lambda_n)} e^{-\lambda_n^2 t}.$$

Hence for any numerical value of h the solution

$$u_N(x,t,h) = w_N(x,t) + v(x)$$

is essentially given by formula so that the error $E(h)$ is readily plotted. To give a numerical demonstration suppose that $\hat{w}(t) \equiv 1$ and that the measured data are taken from the arbitrarily chosen function

$$U(t) = \frac{(1-e^{-t})^4}{2}.$$

Let the experiment be observed over the interval $[0,T]$ and data collected at 200 evenly spaced times $t_m = \frac{mT}{200}$, $m = 1,\ldots,200$. When we compute $E(h)$ for $h = .1*i$, $i = 0,\ldots,50$, with ten terms in the eigenfunction expansion, and then minimize $E(h)$ the following results are obtained for the minimizer h^*:

T	h^*
1	2.6
2	1.5
4	1.2
8	1.1
16	1.1
32	1.0

These results indicate that the assumed boundary temperature $U(t)$ is not consistent with any solution of the model problem for a constant h. But they also show that as $t \to \infty$ and $U(t) \to 1/2$ the numerical results converge to the heat transfer coefficient $h = 1$ consistent with the steady-state solution

$$v(x) = 1 - x/2.$$

This behavior of the computed sequence $\{h^*\}$ simply reflects that more and more data are collected near the steady state as $T \to \infty$.

Example 6.7 Radial heat flow in a sphere.

The next example is characterized by somewhat more complex eigenfunctions than have arisen heretofore.

A sphere of radius R is initially at a uniform temperature $u = 1$. At time $t = 0$ the boundary is cooled instantaneously to and maintained at $u(R,t) = 0$. We wish to find the time required for the temperature at the center of the sphere to fall to $u(0,t) = .5$.

The temperature in the sphere is modeled by the radial heat equation

$$\mathcal{L}u \equiv u_{rr} + \frac{2}{r}u_r - u_r = 0$$

subject to

$$u_r(0, t) = 0, \qquad u(R, t) = 0$$

$$u(r, 0) = 1.$$

This problem already has homogeneous boundary conditions and needs no further transformation. The associated eigenvalue problem is

$$\phi''(r) + \frac{2}{r} \phi'(r) = \mu \phi(r)$$

$$\phi'(0) = \phi(R) = 0.$$

The key observation, found in [12], is that the differential equation can be rewritten as

$$(r\phi(r))'' = \mu (r\phi(r))$$

$$(r\phi(r))(R) = 0.$$

We do not have a boundary condition for $(r\phi(r))$ at $r = 0$ but if we make the reasonable assumption that $\lim_{r \to 0} |\phi(r)| < \infty$, then we have a singular Sturm–Liouville problem and it is readily verified that

$$\phi(r) = \frac{\sin \lambda r}{r}$$

for $\mu = -\lambda^2$ satisfies the differential equation and the boundary conditions

$$\phi(0) = \lambda, \qquad \phi'(0) = 0.$$

The boundary condition at $r = R$ requires that

$$\sin \lambda R = 0$$

so that we have the eigenfunctions

$$\phi_n(r) = \frac{\sin \lambda_n r}{r}, \qquad \lambda_n = \frac{n\pi}{R}, \qquad n = 1, 2 \ldots.$$

By inspection we find that the eigenfunctions $\{\phi_n(r)\}$ are orthogonal in $L_2(0, R, r^2)$. If we now write

$$u_N(r, t) = \sum_{n=1}^{N} \alpha_n(t) \phi_n(r)$$

and substitute it into the radial heat equation we obtain from

$$\sum_{n=1}^{N} \left[-\lambda_n^2 \alpha_n(t) - \alpha_n'(t) \right] \phi_n(r) = 0$$

$$u_N(r,0) = \sum_{n=1}^{N} \frac{\langle 1, \phi_n(r) \rangle}{\langle \phi_n, \phi_n \rangle} \phi_n(r)$$

that

$$u_N(r,t) = \sum_{n=1}^{N} \frac{\langle 1, \phi_n(r) \rangle}{\langle \phi_n, \phi_n \rangle} e^{-\lambda_n^2 t} \phi_n(r).$$

A straightforward calculation shows that

$$\langle 1, \phi_n \rangle = \int_0^R \sin \lambda_n r \, r \, dr = \frac{-R \cos \lambda_n R}{\lambda_n} = \frac{(-1)^{n+1} R}{\lambda_n}$$

$$\langle \phi_n, \phi_n \rangle = \frac{R}{2}$$

so that

$$u_N(r,t) = 2 \sum_{n=1}^{N} (-1)^{n+1} e^{-\lambda_n^2 t} \frac{\sin \lambda_n r}{\lambda_n r}.$$

We observe that

$$u_N(0,0) = 2 \sum_{n=1}^{N} (-1)^{n+1} = \begin{cases} 0 & \text{if } N \text{ is even} \\ 2 & \text{if } N \text{ is odd.} \end{cases}$$

Hence the orthogonal projection u_N does not converge to the initial condition $u_0(r) = 1$ at $r = 0$ as $N \to \infty$. The general theory lets us infer mean square convergence on $(0, R)$. For $r > 0$ and $t = 0$ we do have slow pointwise convergence and for $t > 0$ we have convergence for all $r \in [0, R]$. These comments are illustrated by the data of Table 6.7 computed for $R = 1$.

Table 6.7: Numerical values of $u_N(r,t)$ for radial heat flow on a sphere

N	$u_N(.001, 0)$	$u_N(0, .1)$	$u_N(.001, .1)$
10	.000180	0.707100	.707099
100	.016531	0.707100	.707099
1000	1.000510	0.707100	.707099
10000	.999945	0.707100	.707099

To find the time $\hat{t}$ when $u(0, \hat{t}) = .5$ we need to solve

$$.5 = 2 \sum_{n=1}^{N} (-1)^{(n+1)} z^{n^2}$$

where

$$z = e^{-\left(\frac{\pi}{R}\right)^2 \hat{t}}.$$

The numerical answer is found to be

$$z \cong .25417$$

for all $N > 2$. Hence

$$\hat{t} \cong \frac{|\ln .25417|}{\pi^2} R^2 \cong .1388 R^2.$$

Example 6.8 boundary layer problem.

This example describes a convection dominated diffusion problem. The singular perturbation nature of this problem requires care in how the boundary conditions are made homogeneous in order to obtain a useful analytic solution.

We shall consider heat flow with convection and slow diffusion modeled by

$$\mathcal{L}u \equiv \epsilon u_{xx} + u_x - u_t = 0$$

$$u(0, t) = \sin \omega t, \quad u(1, t) = 0$$

$$u(x, 0) = 0.$$

The maximum principle assures that $|u(x, t)| \le 1$ for all positive ϵ. However, for $\epsilon \ll 1$ the solution can be expected to change rapidly near $x = 0$. u is said to have a boundary layer at $x = 0$.

If $v(x, t)$ is a smooth function such that $v(0, t) = \sin \omega t$ and $v(x, 0) = 0$, then

$$w(x, t) = u(x, t) - v(x, t)$$

satisfies

$$\mathcal{L}w \equiv \epsilon w_{xx} + w_x - w_t = -\epsilon v_{xx} - v_x + v_t \equiv G(x, t)$$

$$w(0, t) = w(1, t) = 0$$

$$w(x, 0) = 0.$$

The associated eigenvalue problem is

$$\epsilon \phi'' + \phi' = \mu \phi$$

$$\phi(0) = \phi(1) = 0.$$

This is a special case of (3.8). The eigenfunctions and eigenvalues of this problem are

$$\phi_n(x) = e^{-\frac{x}{2\epsilon}} \sin \lambda_n x$$

where

$$\lambda_n = n\pi \quad \text{and} \quad \mu_n = -\lambda_n^2 \epsilon - \frac{1}{4\epsilon}.$$

The eigenfunctions are orthogonal with respect to the inner product

$$\langle f, g \rangle = \int_0^1 f(x)g(x)e^{x/\epsilon}dx.$$

For all practical purposes the eigenfunctions vanish outside a boundary layer of order ϵ. It now becomes apparent why our usual choice of

$$v(x,t) = (1-x)\sin\omega t$$

is likely to fail. The right-hand side $G(x,t)$ corresponding to this v is

$$G(x,t) = \sin\omega t + (1-x)\omega\cos\omega t.$$

This function cannot be approximated well in the subspace $M = \text{span}\{\phi_n\}$ for $\epsilon \ll 1$. In view of the discussion to follow in Section 6.3, we shall use instead

$$v(x,t) = (1-x)e^{-\frac{x}{2\epsilon}}\sin\omega t.$$

Then

$$G(x,t) = \frac{1}{4\epsilon}(1-x)e^{-\frac{x}{2\epsilon}}\sin\omega t + \omega(1-x)e^{-\frac{x}{2\epsilon}}\cos\omega t.$$

The approximating problem is

$$\epsilon w_{xx} + w_x - w_t = \sum_{n=1}^{N}\left[\frac{1}{4\epsilon}\hat{\gamma}_n\sin\omega t + \omega\hat{\gamma}_n\cos\omega t\right]\phi_n(x)$$

$$w(0,t) = w(1,t) = w(x,0) = 0,$$

where

$$\hat{\gamma}_n = \frac{\langle(1-x)e^{-\frac{x}{2\epsilon}},\phi_n\rangle}{\langle\phi_n,\phi_n\rangle} = \frac{2}{\lambda_n}.$$

Its solution is

$$w_N(x,t) = \sum_{n=1}^{N}\alpha_n(t)\phi_n(x)$$

where

$$\mu_n\alpha_n - \alpha_n' = A_n\sin\omega t + B_n\cos\omega t$$

with

$$A_n = \frac{1}{2\lambda_n\epsilon}, \qquad B_n = \frac{2\omega}{\lambda_n}.$$

The $\{\alpha_n\}$ are found with the method of undetermined coefficients as

$$\alpha_n(t) = \frac{A_n\mu_n - B_n\omega}{\mu_n^2 + \omega^2}\sin\omega t + \frac{A_n\omega + B_n\mu}{\mu_n^2 + \omega^2}(\cos\omega t - e^{\mu_n t}).$$

It is straightforward to verify that $\alpha_n(t)$ is uniformly bounded with respect to ϵ. The approximate solution of the original problem is

$$u_N(x,t) = \sum_{n=1}^{N}\alpha_n(t)\phi_n(x) + (1-x)e^{-\frac{x}{2\epsilon}}\sin\omega t.$$

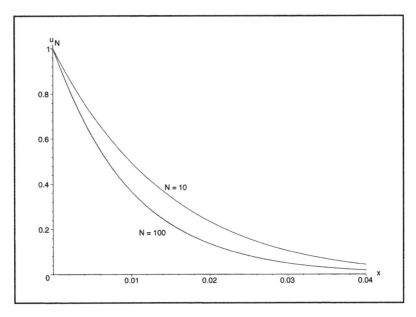

Figure 6.8: Plot of the boundary layer solution $u_N(x, 1.57)$.

Fig. 6.8 shows a plot of the boundary layer of $u_N(x, t)$ for $\epsilon = 0.01$, $x \in [0, .01]$, and $t = 1.57$ when $N = 10$, $N = 90$, and $N = 100$. Since

$$\max_{x \in [0, .01]} |u_{100}(x, 1.57) - u_{90}(x, 1.57)| = 0.0014,$$

the plots for $N = 90$ and $N = 100$ look the same.

Example 6.9 The Black–Scholes equation.

A typical mathematical problem in the pricing of financial options is the following boundary value problem:

$$\frac{1}{2}\sigma^2 x^2 u_{xx} + rx u_x - ru - u_t = 0, \qquad x \in (x0, x1), \quad t \in (0, T]$$

$$u(x0, t) = 0, \qquad u(x1, t) = 0$$

$$u(x, 0) = u_0(x).$$

The differential equation is the so-called Black–Scholes equation. It describes the scaled value $u(x, t)$ of an option with final payoff $u_0(x)$. x denotes the scaled value of the asset on which the option is written and $t = T - \tau$ where τ is real time and T is the time of expiry of the option. The boundaries $x0$ and $x1$ are known as barriers where the option becomes worthless should the value of the underlying asset reach these barriers during its lifetime. σ and r are positive financial parameters.

The associated eigenvalue problem is

$$\frac{1}{2}\sigma^2 x^2 \phi'' + r x \phi' = \mu \phi$$

$$\phi(x0) = \phi(x1) = 0.$$

The discussion of equation (3.8) applies. For $x0 > 0$ we have a regular Sturm–Liouville problem with orthogonal eigenfunctions in the space $\mathcal{L}_2(x0, x1, w)$ where

$$w(x) = x^{\frac{2r}{\sigma^2}-2} = x^{-2A-1}$$

with $A = \frac{1}{2} - \frac{r}{\sigma^2}$. Note that A may be positive or negative. Eigenvalues and eigenfunctions can be found analytically because the eigenvalue equation is a Cauchy–Euler equation with fundamental solutions of the form

$$\phi = x^s$$

where the complex number s is to be determined. It is straightforward to verify that the eigenfunctions and eigenvalues are

$$\phi_n(x) = x^A \sin(\lambda_n \ln(x/x0))$$

$$\lambda_n = \frac{n\pi}{\ln(x1/x0)}$$

$$\mu_n = -\frac{\sigma^2}{2}(A^2 + \lambda_n^2).$$

The approximate solution of the option problem is given by

$$u_N(x,t) = \sum_{n=1}^{N} \alpha_n(t)\phi_n(x)$$

where

$$\mu_n \alpha_n - r\alpha_n - \alpha_n' = 0$$

$$\alpha_n(0) = \hat{a}_n = \frac{\langle u_0, \phi_n \rangle}{\langle \phi_n, \phi_n \rangle}.$$

We observe that if we have a so-called power option

$$u_0(x) = \begin{cases} x^\beta & \text{over } (a,b) \subset (x0, x1) \\ 0 & \text{otherwise} \end{cases}$$

where β is any real number, then

$$\langle u_0, \phi_n \rangle = \int_a^b x^\gamma \sin\left(\lambda_n \ln(x/x0)\right) \frac{dx}{x}$$

where $\gamma = \beta - A$. With the change of variable $y = \frac{\ln(x/x0)}{\ln(x1/x0)}$ the integrals $\langle u_0, \phi_n \rangle$ and $\langle \phi_n, \phi_n \rangle$ can be evaluated analytically. A calculation shows that the Fourier coefficient $\hat{a}_n$ is given by

$$\hat{a}_n = \left. \frac{2x\gamma_0 e^{dy}\left[d\pi \sin n\pi y - n\pi \cos n\pi y\right]}{d^2 + (n\pi)^2} \right|_{y_0}^{y_1}$$

where $d = \gamma \ln(x1/x0)$, $y_1 = \ln(b/x0)/\ln(x1/x0)$, and $y_0 = \ln(a/x0)/\ln(x1/x0)$. For other payoffs $\hat{a}_n$ may have to be found numerically.

The separation of variables solution for this double barrier problem is given by formula

$$u_N(x,t) = \sum_{n=1}^{N} \hat{a}_n e^{(\mu_n - r)t} \phi_n(x).$$

It is known that many barrier problems for power options have an analytic solution in terms of error functions so that our separation of variables solution can be compared with the true solution. A comparison of our approximate solution with the analytic solution of Haug [9] for an option known as an "up and out call" provides insight into the accuracy of the approximate solution. The up and out call is characterized by the boundary and initial data

$$u(0,t) = u(x1,t) = 0, \qquad x1 > 1$$

$$u_0(x) = \max\{0, x - 1\}.$$

For our eigenfunction approach the boundary condition at $x = 0$ must be replaced by $u(x0,t) = 0$ for some $x0 > 0$.

Since the boundary/initial data are discontinuous at $x1$, many terms of the series are required for small t to cope with the Gibbs phenomenon in the approximation of $u_0(x)$. For example, a comparison with the error function solution shows that for $r = .04$, $\sigma = .3$, $x0 = .001$, and $x1 = 1.2$ we need at $t = .00274$ (= one day before expiration) about 400 terms for financially significant accuracy while for $t = .25$ (= 3 months) 50 terms suffice for the same accuracy.

It appears difficult to prove that the Fourier series solution converges to the error function solution as $N \to \infty$ and $x0 \to 0$; however, it is readily established that for this payoff the approximate solution stays bounded as $x0 \to 0$. The computations show that moving the boundary condition to an artificial barrier at $x = x0 > 0$ as required for the eigenfunction approach has little influence on the results.

As another application we show in Fig. 6.9 the price of a so-called double barrier straddle characterized by the payoff

$$u_0(x) = \max\{1 - x, 0\} + \max\{x - 1, 0\}$$

with a down and out barrier at $x0 = .8$ and an up and out barrier at $x1 = 1.2$ so that

$$u(.8, t) = u(1.2, t) = 0.$$

Here the boundary/initial data are discontinuous at $x0$ and $x1$. The solution is shown at $t = .00274$ for $N = 400$ and at $t = .25$ for $N = 50$.

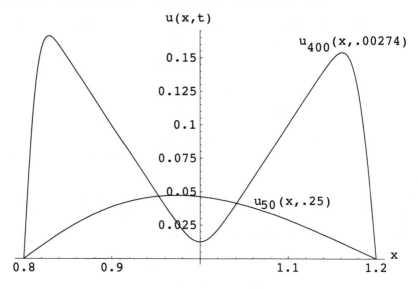

Figure 6.9: Scaled price of a "double barrier straddle" option one day ($t = .00274$) and three months ($t = .25$) before expiration. $r = .04$, $\sigma = .3$.

Example 6.10 Radial heat flow in a disk.

Let us now turn to the analogue of Example 6.6 and consider heat flow in a disk. This problem is more complicated than the flow in a sphere and will be our first introduction to Bessel functions.

A disk of radius R is initially at a uniform temperature $u_0 = 1$. At time $t = 0$ the boundary is cooled instantaneously to and maintained at $u(R, t) = 0$. We want the time required for the temperature at the center of the disk to fall to $u(0, t) = .5$.

Since there is no angular dependence in the data, the temperature $u(r, t)$ is given by the radial heat equation

$$\mathcal{L}u \equiv u_{rr} + \frac{1}{r} u_r - u_t = 0$$

subject to the symmetry condition

$$u_r(0, t) = 0$$

and the initial and boundary data

$$u(R,t) = 0, \qquad u(r,0) = 1.$$

Since the boundary data already are homogeneous, we see that the eigenvalue problem associated with the spatial part of the radial heat equation is

$$\phi''(r) + \frac{1}{r}\phi'(r) = \mu\phi(r)$$

$$\phi'(0) = 0, \qquad \phi(R) = 0.$$

The equations can be transformed to standard form as described in Chapter 3

$$(r\phi'(r))' = \mu r\phi(r) \tag{6.4}$$

$$\phi'(0) = 0, \qquad \phi(R) = 0.$$

If this problem were given on an annulus $r_0 < r < R$ with $r_0 > 0$, then it would be a standard Sturm–Liouville problem with countably many eigenvalues and eigenfunctions, and with eigenfunctions for distinct eigenvalues orthogonal in $L_2(r_0, R, r)$.

The general theory does not apply because the coefficient of $\phi''(r)$ vanishes at $r = 0$. This makes the problem a singular Sturm–Liouville problem. Fortunately, the conclusions of the general theory remain applicable. Equation (6.4) is a special form of Bessel's equation and can be matched with (3.10). It has negative eigenvalues so that we can write

$$-\mu = \lambda^2.$$

For arbitrary λ the solution of Bessel's equation satisfying $\phi'(0) = 0$ is the so-called Bessel function of the first kind of order zero given by

$$J_0(\lambda r) = \sum_{m=0}^{\infty} \frac{(-1)^m}{m!m!} \left(\frac{\lambda r}{2}\right)^{2m}.$$

A plot of $J_0(x)$ vs. x is shown in Fig. 6.10.

Like $\cos \lambda r$ the Bessel function oscillates and the zero-crossings depend on λ. Different eigenfunctions are found if λ_n is chosen such that

$$J_0(\lambda_n R) = 0.$$

It follows that there are countably many eigenvalues $0 < \lambda_1 < \lambda_2 < \cdots$ where

$$\lambda_n R = x_{n0}$$

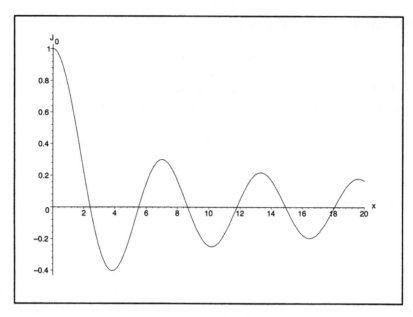

Figure 6.10: Plot of $J_0(x)$ vs. x.

is the nth root of the Bessel function $J_0(x)$. These roots are available from Maple, etc. and will be considered known. Finally, since $J_0(0) = 1$ and $J_0'(0) = 0$, it is straightforward to verify as in the regular Sturm–Liouville case that

$$\int_0^R J_0(\lambda_m r) J_0(\lambda_n r) r\, dr = 0, \qquad m \neq n,$$

i.e., that the eigenfunctions corresponding to distinct eigenvalues are orthogonal in $L_2(0, R, r)$.

We now find an approximate solution of the heat flow problem in the usual way. We solve

$$u_{rr} + \frac{1}{r} u_r - u_t = 0$$

$$u_r(0, t) = u(R, t) = 0$$

with the projected initial condition

$$P_N u(r, 0) = \sum_{n=1}^N \hat{\alpha}_n \phi_n(r)$$

where

$$\hat{\alpha}_n = \frac{\langle 1, \phi_n(r) \rangle}{\langle \phi_n, \phi_n \rangle}.$$

In this case

$$\phi_n(r) = J_0(\lambda_n r) \quad \text{and} \quad \langle f(r), g(r) \rangle = \int_0^R f(r)g(r)rdr.$$

The solution of this problem is

$$u_N(r,t) = \sum_{n=1}^N \alpha_n(t)\phi_n(r)$$

where

$$-\lambda_n^2 \alpha(t) - \alpha_n'(t) = 0$$

$$\alpha_n(0) = \hat{\alpha}_n.$$

Hence

$$u_N(r,t) = \sum_{n=1}^N \hat{\alpha}_n e^{-\lambda_n^2 t} J_0(\lambda_n r).$$

The evaluation of the inner products involving Bessel functions is not quite as forbidding for this model problem as might appear from the series definition of the Bessel function. Numerous differential and integral identities are known for Bessel functions of various orders. For example, it can be shown that

$$\langle 1, J_0(\lambda_n r) \rangle = \frac{R^2}{x_n} J_1(x_{n0})$$

$$\langle J_0(\lambda_n r), J_0(\lambda_n r) \rangle = \frac{R^2}{2} J_1^2(x_{n0})$$

where x_{n0} is the nth root of $J_0(x) = 0$ and $J_1(x)$ is the Bessel function of order 1 which also is tabulated or available from computer libraries. Using the values given in [19, p. 261] we find

n	x_n	$J_1(x_n)$	$\hat{\alpha}_n$
1	2.405	.5191	1.602
2	5.520	−.3403	−1.065
3	8.654	.2715	0.8512
4	11.792	−.2325	−0.7295

If we set

$$z = e^{-\frac{t}{R^2}},$$

then the approximate solution to our problem is that value of z which satisfies

$$.5 = \sum_{n=1}^N \hat{\alpha}_n z^{x_{n0}^2}.$$

For $N = 2, 3$, and 4 the computer yields

$$z \cong .818$$

so that the temperature at the center of the disk is reduced to half its original value at time

$$\hat{t} \cong R^2 |\ln .818| \cong .2009 R^2.$$

Example 6.11 Heat flow in a composite slab.

This example requires the eigenfunctions of the interface problem discussed in Chapter 3.

A composite slab of thickness L consists of material with diffusivities α_1 and α_2 and conductivities k_1 and k_2 for $x \in (0, X)$ and $x \in (X, L)$, respectively, where $X \in (0, L)$ is a given interface between the two materials. We are again interested in the phase shift of the thermal wave passing through the slab as a function of the problem parameters.

The mathematical model is

$$\mathcal{L}_1 u_1 \equiv \alpha_1 u_{1xx} - u_{1t} = 0, \qquad x \in (0, X), \quad t > 0$$

$$\mathcal{L}_2 u_2 \equiv \alpha_2 u_{2xx} - u_{2t} = 0, \qquad x \in (X, L), \quad t > 0.$$

We use the boundary and initial conditions of Example 6.4

$$u_1(0, t) = \sin \omega t, \quad u_{2x}(L, t) = 0$$

$$u_1(x, 0) = u_2(x, 0) = 0.$$

Continuity of temperature and heat flux at $x = X$ require

$$u_1(X, t) = u_2(X, t)$$

$$k_1 u_{1x}(X, t) = k_2 u_{2x}(X, t).$$

This problem is converted to one with homogeneous boundary data by setting

$$w_i(x, t) = u_i(x, t) - \sin \omega t, \quad i = 1, 2.$$

Then

$$\mathcal{L}_1 w_1 = \omega \cos \omega t, \quad \mathcal{L}_2 w_2 = \omega \cos \omega t$$

$$w_1(0, t) = w_{2x}(L, t) = 0$$

$$w_1(x, 0) = w_2(x, 0) = 0.$$

The interface conditions remain

$$w_1(X, t) = w_2(X, t)$$

$$k_1 w_{1x}(X, t) = k_2 w_{2x}(X, t).$$

The associated eigenvalue problem is

$$\phi'' = \mu\phi, \quad x \in (0, X)$$

$$\phi'' = \mu\phi, \quad x \in (X, L)$$

$$\phi(0) = \phi'(L) = 0$$

$$\phi(X-) = \phi(X+), \quad k_1\phi'(X-) = k_2\phi'(X+).$$

This is precisely the eigenvalue problem discussed at the end of Chapter 3. The eigenfunction satisfying the boundary conditions is of the form

$$\phi(x) = c\sin\frac{\lambda x}{\sqrt{\alpha_1}}, \quad x \in (0, X)$$

$$\phi(x) = d\cos\frac{\lambda(L-x)}{\sqrt{\alpha_2}}, \quad x \in (X, L)$$

where

$$\mu = -\lambda^2.$$

The interface conditions can be written in matrix form

$$\begin{pmatrix} \sin\frac{\lambda X}{\sqrt{\alpha_1}} & -\cos\frac{\lambda(L-X)}{\sqrt{\alpha_2}} \\ \frac{k_1}{\sqrt{\alpha_1}}\cos\frac{\lambda X}{\sqrt{\alpha_1}} & -\frac{k_2}{\sqrt{\alpha_2}}\sin\frac{\lambda(L-X)}{\sqrt{\alpha_2}} \end{pmatrix} \begin{pmatrix} c \\ d \end{pmatrix} = \begin{pmatrix} 0 \\ 0 \end{pmatrix}.$$

The admissible values of λ are the roots of

$$f(\lambda) \equiv \frac{k_2}{\sqrt{\alpha_2}}\sin\frac{\lambda(L-X)}{\sqrt{\alpha_2}}\sin\frac{\lambda X}{\sqrt{\alpha_1}} - \frac{k_1}{\sqrt{\alpha_1}}\cos\frac{\lambda(L-X)}{\sqrt{\alpha_2}}\cos\frac{\lambda X}{\sqrt{\alpha_1}} = 0.$$

Thus λ must solve the equation

$$\tan\frac{\lambda(L-X)}{\sqrt{\alpha_2}}\tan\frac{\lambda X}{\sqrt{\alpha_1}} = \frac{k_1\sqrt{\alpha_2}}{k_2\sqrt{\alpha_1}}. \tag{6.5}$$

To be consistent with the notation of Example 6.5 let us index the roots of equation (6.5) such that

$$0 < \lambda_0 < \lambda_1 < \cdots.$$

The corresponding eigenfunctions are

$$\phi_n(x) = \begin{cases} \cos\frac{\lambda_n(L-X)}{\sqrt{\alpha_2}}\sin\frac{\lambda_n x}{\sqrt{\alpha_1}}, & x \in (0, X) \\ \sin\frac{\lambda_n X}{\sqrt{\alpha_1}}\cos\frac{\lambda_n(L-x)}{\sqrt{\alpha_2}}, & x \in (X, L). \end{cases}$$

The remainder of the problem is identical to that of Example 6.5. The problem is

$$\mathcal{L}w = \omega \cos \omega t.$$

The approximation is

$$\mathcal{L}w = \omega \cos \omega t \sum_{n=0}^{N} \hat{\gamma}_n \phi_n(x)$$

where

$$\hat{\gamma}_n = \frac{\langle 1, \phi_n \rangle}{\langle \phi_n, \phi_n \rangle}$$

with

$$\langle f, g \rangle = \int_0^X f(x)g(x)k_1\alpha_2 dx + \int_X^L f(x)g(x)k_2\alpha_1 dx.$$

The solution is

$$u_N(x,t) = \sum_{n=0}^{N} \alpha_n(t)\phi_n(x) + \sin \omega t$$

where

$$-\lambda_n^2 \alpha_n - \alpha_n' = \hat{\gamma}_n \omega \cos \omega t.$$

It follows as in Example 6.5 that $u_N(x,t)$ is given by

$$u_N(L,t) = \sum_{n=0}^{N} -\frac{\hat{\gamma}_n \lambda_n^2}{\sqrt{\lambda_n^4 + \omega^2}} \sin(\omega t - \psi_n)\phi_n(L)$$

where

$$\sin \psi_n = \frac{\omega}{\sqrt{\lambda_2^4 + \omega^2}}.$$

The dominant phase shift again corresponds to $n = 0$.

It is straightforward to show that the phase shift for this composite slab reduces to that of the homogeneous slab found in Example 6.5 if the thermal parameters of both components are the same (see Exercise 6.3).

Example 6.12 Reaction-diffusion with blowup.

The eigenfunction expansion approach is applicable to some nonlinear reaction diffusion problems, although in practice it will have to be combined with a numerical method to solve the differential equations for the $\{\alpha_n(t)\}$. This makes our separation of variables technique a spectral method in the numerical analysis sense. In order to indicate the connection we shall apply the eigenfunction expansion method to the nonlinear equation

$$\mathcal{L}u = u_{xx} - u_t = ku^2, \qquad k > 0,$$

which is representative for diffusion with polynomial source terms. We impose the boundary and initial condition

$$u(0,t) = u(1,t) = 0$$
$$u(x,0) = u_0(x)$$

where u_0 is given. The approximating problem is again written as

$$\mathcal{L}u = u_{xx} - u_t = P_N u^2$$

$$u(0,t) = u(1,t) = 0$$
$$u(x,0) = P_N u_0(x)$$

where P_N denotes the orthogonal projection in $\mathcal{L}_2(0,1)$ into the usual subspace

$$M_N = \text{span}\{\phi_n\}_{n=1}^N$$

of the eigenfunctions $\phi_n(x) = \sin \lambda_n x$, $\lambda_n = n\pi$. It is a question of mathematical analysis whether this problem has a unique solution. Here we want to demonstrate that the problem admits a unique eigenfunction solution of the familiar form

$$u_N(x,t) = \sum_{n=1}^N \alpha_n(t)\phi_n(x),$$

for $0 < t < T$ for some $T > 0$.

When we substitute u_N into the differential equation and equate the terms multiplying $\phi_n(x)$, we obtain for $n = 1, 2, \ldots, N$

$$-\lambda_n^2 \alpha_n(t) - \alpha_n'(t) = 2k \left[\sum_{i=1}^N \sum_{j=1}^N \alpha_i(t)\alpha_j(t)\langle \phi_i\phi_j, \phi_n \rangle \right]$$

$$\alpha_n(0) = 2\langle u_0, \phi_n \rangle.$$

This is an initial value problem for a system of N nonlinear first order ordinary differential equations of the form

$$\vec{\alpha}' = F(t, \vec{\alpha}).$$

By inspection we see that $F_t(t, \vec{\alpha}) = 0$ and that F is differentiable with respect to the components of $\vec{\alpha}$. The theory of ordinary differential equations guarantees existence and uniqueness of a solution $\vec{\alpha}(t)$ as long as the components of $\vec{\alpha}$ remain bounded.

In contrast to most of the application given above it appears no longer possible to find $\vec{\alpha}(t)$ analytically. When numerical methods are applied to integrate the equations for $\{\alpha_n(t)\}$, new issues arise with the accuracy and stability of the computation. Such concerns are addressed in studies of numerical spectral methods based on expansions u_N where the $\{\phi_n\}$ are not necessarily eigenfunctions (for a recent reference see, e.g., [8].)

For an illustration let us find an approximate solution when

$$u_0(x) = 4x(1 - x)(1 - 4x).$$

This initial value was chosen because we can expect decay of the solution to $u \equiv 0$ for small k and blowup (to $-\infty$) if k is large.

Because we merely wish to demonstrate the possibility of attacking nonlinear systems with an eigenfunction expansion, we restrict our study to $N \leq 4$ and we make no claim about convergence as N increases. Representative numerical results are shown in Fig. 6.12 for $N = 3$. Plotted are the solution $u_3(.1, t)$ for $k = 10.46$ which decays to zero as t increases, and the solution $u_3(.1, t)$ for $k = 10.47$ which blows up for $t > .7$. Blowup occurs whenever the (black box) Maple integrator fails to integrate the nonlinear system for the $\{\alpha_n(t)\}$ beyond some value $T > 0$. The threshhold value for k between decay and blowup as a function of N is observed to be: $k(1) \cong 11.26$, $k(2) \cong 10.50$, $k(3) = k(4) \cong 10.46$.

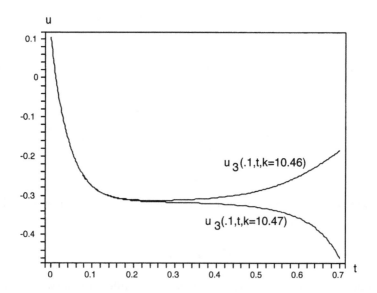

Figure 6.12: Decaying solution $u_3(.1, t, k = 10.46)$ and unbounded solution $u_3(.1, t, k = 10.47)$ of $u_{xx} - u_t = ku^2$.

6.2 Convergence of $u_N(x,t)$ to the analytic solution

The dominant question for our approximation method has to be: How is the computed approximation $u_N(x,t)$ related to the analytic solution?

This question was considered in Example 6.1 by assuming that the analytic solution is given by $u(x,t) = \lim_{N\to\infty} u_N(x,t)$ and estimating the error incurred by truncating the infinite series after N terms. Consequently, the answer is very specific for the model problem of Example 6.1. Here we give answers which rely only on the existence of an analytic solution and its smoothness properties as known from the theory of partial differential equations.

For ease of exposition we shall consider the model problem

$$\mathcal{L}u \equiv u_{xx} - u_t = F(x,t) \qquad (6.6)$$

$$u(0,t) = A(t), \qquad u(L,t) = B(t)$$

$$u(x,0) = u_0(x).$$

If $v(x,t)$ is a smooth function used to subtract out the boundary conditions, then we see from

$$u(x,t) = w(x,t) + v(x,t)$$

and

$$u_N(x,t) = w_N(x,t) + v(x,t)$$

that convergence of u_N to u is equivalent to convergence of w_N to w where

$$\mathcal{L}w \equiv w_{xx} - w_t = F(x,t) - \mathcal{L}v \equiv G(x,t) \qquad (6.7)$$

$$w(0,t) = w(L,t) = 0$$

$$w(x,0) = u_0(x) - v(x,0) \equiv w_0(x)$$

and w_N is the approximate solution of (6.7) when G and w_0 are replaced by $P_N G$ and $P_N w_0$.

We shall assume that $w_0 \in L_2(0,L)$, that G is continuous on $(0,L) \times (0,T]$, that $G(x,t) \in L_2(0,L)$ for each t, and that $\|G(x,t)\|$ is bounded in t. These conditions are sufficient to guarantee that (6.7) has a solution w such that w_x and w_t are square integrable over $(0,L)$ for all t (see, e.g., [5]). We now employ energy estimates common in the finite element method to establish mean square convergence of w_N to w. Let

$$e(x,t) = w(x,t) - w_N(x,t)$$

denote the error of the approximation. Then $e(x, t)$ is the solution of

$$\mathcal{L}e \equiv e_{xx} - e_t = G(x, t) - P_N G(x, t) \tag{6.8}$$

$$e(0, t) = e(L, t) = 0$$

$$e(x, 0) = w_0(x) - P_N w_0(x).$$

If we multiply the differential equation by $e(x, t)$ and integrate with respect to x, we obtain

$$\int_0^L (e_{xx} - e_t) e \, dx = \int_0^L [G - P_N G] e \, dx.$$

After integrating the first term by parts the following equation results:

$$\frac{1}{2} \frac{d}{dt} \int_0^L e^2(x, t) dx = - \int_0^L e_x(x, t)^2 dx - \int_0^L [G - P_N G] \, e(x, t) dx.$$

Example 4.20 proves that

$$\int_0^L e(x, t)^2 dx \leq \left(\frac{L}{\pi} \right)^2 \int_0^L e_x(x, t)^2 dx.$$

This inequality allows the following estimate for the mean square error at time t:

$$\frac{d}{dt} E(t) \leq -2 \left(\frac{\pi}{L} \right)^2 E(t) + 2 \int_0^L |G(x, t) - P_N G(x, t)| |e(x, t)| dx \tag{6.9}$$

where for convenience we have set

$$E(t) = \int_0^L e(x, t)^2 dx.$$

Applying the algebraic–geometric mean inequality

$$2 \left(\frac{a}{\sqrt{\epsilon}} \sqrt{\epsilon} \, b \right) \leq \frac{a^2}{\epsilon} + \epsilon b^2$$

with $\epsilon = \left(\frac{\pi}{L} \right)^2$ we obtain the estimate

$$\int_0^L [G(x, t) - P_N G(x, t)] \, e(x, t) dx$$

$$\leq \left(\frac{L}{\pi} \right)^2 \int_0^L [G(x, t) - P_N G(x, t)]^2 \, dx + \left(\frac{\pi}{L} \right)^2 E(t).$$

With this estimate for (6.9) we have the following error bound:

$$E'(t) \leq - \left(\frac{\pi}{L} \right)^2 E(t) + \left(\frac{L}{\pi} \right)^2 \|G - P_N G\|^2 \tag{6.10}$$

$$E(0) = \|w_0 - P_N w_0\|^2$$

where $\| \ \|$ is the usual norm of $L_2(0, L)$. Inequalities like (6.10) occur frequently in the qualitative study of ordinary differential equations. If we express it as an equality

$$E'(t) = -\left(\frac{\pi}{L}\right)^2 E(t) + \left(\frac{L}{\pi}\right)^2 \|G - P_N G\|^2 - g(t)$$

for some nonnegative (but unknown) function $g(t)$, then this equation has the solution

$$E(t) = \|w_0 - P_N w_0\|^2 \, e^{-(\pi/L)^2 t}$$
$$+ \int_0^t e^{-(\pi/L)^2 (t-s)} \left[\left(\frac{L}{\pi}\right)^2 \|G(x,s) - P_N G(x,s)\|^2 - g(s) \right] ds$$

or finally

$$E(t) \le \|w_0 - P_N w_0\|^2 \, e^{-(\pi/L)^2 t}$$
$$+ \left(\frac{L}{\pi}\right)^2 \int_0^t e^{-(\pi/L)^2 (t-s)} \|G(x,s) - P_N G(x,s)\|^2 ds. \qquad (6.11)$$

This inequality is known as Gronwall's inequality for (6.10). Thus the error due to projecting the initial condition depends entirely on how well w_0 can be approximated in $\mathrm{span}\{\phi_n\}_{n=1}^N$ and can be made as small as desirable by taking sufficiently many eigenfunctions. In addition, this contribution to the overall error decays with time. The approximation of the source term G also converges in the mean square sense. If we can assert that $\|G(x,t) - P_N G(x,t)\| \to 0$ uniformly in t, then we can conclude that $E(t) \to 0$ uniformly in t as $N \to \infty$.

As an illustration consider

$$\mathcal{L}w \equiv w_{xx} - w_t = 0$$

$$w(0,t) = w(L,t) = 0$$

$$w(x,0) = 1.$$

It is straightforward to compute that

$$P_N(1) = \frac{4}{\pi} \sum_{n=0}^N \frac{1}{2n+1} \sin\left(\frac{2n+1}{L}\pi x\right).$$

Using Parseval's identity

$$\|1\|^2 = L \frac{8}{\pi^2} \sum_{n=0}^{\infty} \frac{1}{(2n+1)^2}$$

we obtain the estimate

$$\|1 - P_N(1)\|^2 = L\frac{8}{\pi^2} \sum_{n=N+1}^{\infty} \frac{1}{(2n+1)^2} \le \frac{8L}{\pi^2(4N+6)}$$

where the last inequality follows from the integral test

$$\sum_{n=N+1}^{\infty} \frac{1}{(2n+1)^2} \le \int_{N+1}^{\infty} \frac{dx}{(2x+1)^2} = \frac{1}{4N+6} .$$

The mean square error is $E(t) \le \frac{8L}{\pi^2(4N+6)} e^{-\left(\frac{\pi}{L}\right)^2 t}$.

In general we cannot expect much more from our approximate solution because initial and boundary data may not be consistent so that the Gibbs phenomenon precludes a uniform convergence of w_N to w. However, as we saw in Chapter 4, when the data are smooth, then their Fourier series will converge uniformly. In this case it is possible to establish uniform convergence with the so-called maximum principle for the heat equation. Hence let us assume that $w_0(x)$ and $G(x,t)$ are such that

$$\max_x |w_0(x) - P_N w_0(x)| \to 0 \qquad \text{for } x \in [0, L]$$

and

$$\max_{x,t} |G(x,t) - P_N G(x,t)| \to 0 \qquad \text{for } (x,t) \in [0, L] \times [0, T]$$

as $N \to \infty$. Here T is considered arbitrary but fixed. Then the error $e(x,t)$ satisfies (6.8) with continuous initial/boundary data and a smooth source term. The inequality (1.10) of Section 1.4 translates into the following error estimate for this one-dimensional problem:

$$|e(x,t)| \le K \max_{x,t} |G(x,t) - P_N G(x,t)| + \max_x |w_0(x) - P_N w_0(x,t)|$$

where the constant K depends on the length of the interval. In other words, if the orthogonal projections converge uniformly to the data functions, then the approximate solution likewise will converge uniformly to the true solution on the computational domain $[0, L] \times [0, T]$.

We conclude our discussion of convergence with the following characterization of w_N.

Theorem 6.13 *Let w be the analytic solution of (6.7). If for $t > 0$ the derivatives w_{xx} and $w_t \in L_2(0, L)$, then*

$$P_N w(x,t) = w_N(x,t).$$

Proof. Since

$$w_{xx}(x,t) - w_t(x,t) = G(x,t)$$

$$w(0,t) = w(L,t) = 0$$

$$w(x,0) = w_0(x),$$

we see that

$$P_N(w_{xx} - w_t) = P_N G$$

$$P_N w(x,0) = P_N w_0(x).$$

Writing out the projections we obtain for each n

$$\langle w_{xx}, \phi_n \rangle - \langle w_t, \phi_n \rangle = \langle G, \phi_n \rangle.$$

Integration by parts shows that

$$\langle w_{xx}, \phi_n \rangle = -\lambda_n^2 \langle w, \phi_n \rangle,$$

and of course

$$\langle w_t, \phi_n \rangle = \frac{d}{dt} \langle w, \phi_n \rangle.$$

Hence the term $\frac{\langle w, \phi_n \rangle}{\langle \phi_n, \phi_n \rangle}$ satisfies the initial value problem for α_n. Since its solution is unique, it follows that $\alpha_n(t) \equiv \frac{\langle w, \phi_n \rangle}{\langle \phi_n, \phi_n \rangle}$ and hence that

$$w_N(x,t) = P_N w(x,t).$$

Theorem 6.13 states that the computed w_N for any t is the orthogonal projection of the unknown exact solution w onto the span of the first eigenfunctions $\{\phi_n\}$; in other words, w_N is the best possible approximation to w in span $\{\phi_n\}$.

Conditions on the data for the existence of w_{xx} and $w_t \in L_2(0, L)$ are discussed, for example, in [5]. In particular, it is necessary that the consistency conditions $w_0(0) = w_0(L) = 0$ hold. This is a severe restriction on the data and not satisfied by many of our examples.

6.3 Influence of the boundary conditions and Duhamel's solution

The formulas derived for the solution of (6.6) involve the function v used to zero out the boundary conditions. Since there are many v which may be used, and since the analytic solution is uniquely determined by the boundary data and is independent of v, it may be instructive to see how v actually enters the computational solution. We recall the original problem is transformed into

$$\mathcal{L}w = w_{xx} - w_t = \mathcal{L}u - \mathcal{L}v = F - (v_{xx} - v_t)$$

$$w(0,t) = w(L,t) = 0$$

$$w(x,0) = u_0(x) - v(x,0).$$

Let us write the approximate solution $w_N(x,t)$ in the form

$$w_N(x,t) = w_N^1(x,t) + w_N^2(x,t)$$

where

$$\mathcal{L}w_N^1 = P_N F(x,t)$$

$$w_N^1(x,0) = 0$$

and where w_N^2 accounts for the influence of the initial and boundary conditions

$$\mathcal{L}w_N^2 = \sum_{n=1}^{N} -\frac{\langle v_{xx} - v_t, \phi_n \rangle}{\langle \phi_n, \phi_n \rangle} \phi_n(x)$$

$$w_N^2(x,0) = \sum_{n=1}^{N} \frac{\langle u_0(x) - v(x,0), \phi_n \rangle}{\langle \phi_n, \phi_n \rangle} \phi_n(x).$$

The above discussion of the Dirichlet problem shows that

$$w_N^1(x,t) = -\sum_{n=1}^{N} \int_0^t e^{-\lambda_n^2(t-s)} \gamma_n(s) ds \phi_n(x)$$

where

$$\gamma_n(t) = \frac{\langle F(x,t), \phi_n \rangle}{\langle \phi_n, \phi_n \rangle}.$$

We remark that $w_N^1(x,t)$ is identical to the solution obtained with Duhamel's principle since by inspection the function

$$W_n(x,t,s) \equiv e^{-\lambda_n^2(t-s)} \gamma_n(s) \phi_n(x)$$

solves the problem

$$W_{xx} - W_t = 0$$

$$W(0,t,s) = W(L,t,s) = 0$$

$$W(x,s,s) = -\frac{\langle F(x,s), \phi_n \rangle}{\langle \phi_n, \phi_n \rangle}$$

so that

$$\sum_{n=1}^{N} W_n(x, s, s) = -P_N F(x, s)$$

and

$$w_N^1(x, t) = \int_0^t \sum_{n=1}^{N} W_n(x, t, s)ds.$$

In other words, the eigenfunction expansion is identical to the solution obtained with Duhamel's superposition principle.

We shall now compute w_N^2. Integration by parts shows that

$$\langle v_{xx}, \phi_n \rangle = [v_x(x, t)\phi_n(x) - v(x, t)\phi_n'(x)]\Big|_0^L + \langle v, \phi_n'' \rangle = C_n(t) - \lambda_n^2 \langle v, \phi_n \rangle$$

where for the data of (6.6)

$$C_n(t) = \lambda_n[A(t) - B(t) \cos \lambda_n L].$$

$C_n(t)$ is independent of the choice of v since only the boundary data appear. If we write

$$w_N^2(x, t) = \sum_{n=1}^{N} \alpha_n(t)\phi_n(x),$$

then

$$-\lambda_n^2 \alpha_n(t) - \alpha_n'(t) = \frac{1}{\langle \phi_n, \phi_n \rangle} \left[-C_n(t) + \frac{d}{dt} \langle v, \phi_n \rangle + \lambda_n^2 \langle v, \phi_n \rangle \right].$$

It follows that

$$-\lambda_n^2 \left[\alpha_n(t) + \frac{\langle v(x, t), \phi_n \rangle}{\langle \phi_n, \phi_n \rangle} \right] - \frac{d}{dt} \left[\alpha_n(t) + \frac{\langle v(x, t), \phi_n \rangle}{\langle \phi_n, \phi_n \rangle} \right] = \frac{C_n(t)}{\langle \phi_n, \phi_n \rangle}$$

$$\alpha_n(0) + \frac{\langle v(x, 0), \phi_n \rangle}{\langle \phi_n, \phi_n \rangle} = \frac{\langle u_0(x), \phi_n \rangle}{\langle \phi_n, \phi_n \rangle}$$

so that

$$\alpha_n(t) + \frac{\langle v(x, t), \phi_n \rangle}{\langle \phi_n, \phi_n \rangle} = \frac{\langle u_0, \phi_n \rangle}{\langle \phi_n, \phi_n \rangle} e^{-\lambda_n^2 t} + \int_0^t e^{-\lambda_n^2(t-s)} \frac{C_n(s)}{\langle \phi_n, \phi_n \rangle} ds.$$

Thus

$$u_N(x, t) = w_N^1(x, t) + \sum_{n=1}^{N} D_n(t)\phi_n(x) + [v(x, t) - P_N v(x, t)]$$

where

$$D_n(t) = \frac{\langle u_0, \phi_n \rangle}{\langle \phi_n, \phi_n \rangle} e^{-\lambda_n^2 t} + \int_0^t e^{-\lambda_n^2(t-s)} \frac{C_n(s)}{\langle \phi_n, \phi_n \rangle} ds$$

is again independent of the choice of v.

Hence the solution $u_N(x,t)$ at time t depends on the data of the original problem and on the difference between v and its orthogonal projection into the span of the eigenfunctions. In all of our examples any function v with continuous v_{xx} and v_t on $[0,L] \times [0,T]$ will be admissible to zero out nonhomogeneous boundary conditions. Functions linear (or possibly quadratic) in x will cause the fewest technical complications for the eigenfunctions of Table 3.1. However, for the boundary layer problem of Example 6.8 the linear function

$$v(x) = 1 - x$$

is not advantageous. We recall that the eigenfunctions of this example are

$$\phi_n(x) = e^{-\frac{x}{2\epsilon}} \sin \lambda_n x, \qquad \lambda_n = n\pi, \qquad n = 1,2,\dots$$

which are complete and orthogonal in $L_2(0,1,e^{x/\epsilon})$. In this case

$$P_N v(x) = \sum_{n=1}^{N} \hat{\alpha}_n \phi_n(x)$$

where

$$\hat{\alpha}_n = 2 \int_0^1 (1-x) \sin \lambda_n x e^{\frac{x}{2\epsilon}} dx$$
$$= 2 \frac{4\epsilon^2 \lambda_n (1 + 4\epsilon + 4\epsilon^2 \lambda_n^2 + (-1)^{n+1} 4\epsilon e^{1/2\epsilon})}{(1 + 4\epsilon^2 \lambda_n^2)^2}.$$

It is apparent that for $\epsilon \ll 1$ the Fourier coefficients are of order $(-1)^{n+1} e^{1/2\epsilon}/(\epsilon \lambda_n^3)$ which makes the projection $P_N v(x)$ impossible to compute numerically because we have to sum large alternating terms. On the other hand, if we choose

$$v(x) = (1-x)e^{-x/2\epsilon},$$

then

$$P_N v(x) = 2 \sum_{n=1}^{N} \frac{1}{n\pi} \phi_n(x)$$

which is, of course, manageable and allows us to resolve the boundary layer of Example 6.8.

Exercises

6.1) Solve
$$\mathcal{L}u = u_{xx} - u_t = 0$$

$$u(0,t) = te^{-t}, \qquad u_x(1,t) = 0$$
$$u(x,0) = 0.$$

i) Find t^* when $u(1,t)$ reaches its maximum.

ii) Find t^{**} when $u_t(1,t)$ reaches its maximum.

6.2) Solve
$$\mathcal{L}u \equiv u_{xx} - u_t = 0$$

$$u(0,t) = te^{-t}, \qquad u(1,t) = 0$$
$$u(x,0) = 0.$$

i) Find t^* when $|u_x(1,t)|$ reaches its maximum.

ii) Find t^{**} when $|u_{xt}(1,t)|$ reaches its maximum.

6.3) Verify that the solution for the composite slab in Example 6.11 is independent of X for $\alpha_1 = \alpha_2$ and $k_1 = k_2$.

6.4) Show that the problem

i)
$$\mathcal{L}u \equiv ku_{xx} - \rho c u_t = 0, \quad a < x < b, \quad t > 0$$

$$u(a,t) = A, \qquad u(b,t) = B$$
$$u(x,0) = C$$
$$\text{with } A = B$$

can be transformed (i.e., made nondimensional) into

ii)
$$\mathcal{L}U \equiv U_{yy} - U_\tau = 0$$

$$U(0,\tau) = 1, \qquad U(1,\tau) = 0$$
$$U(y,0) = D$$

where D depends on the parameters of problem i).

Solve problem ii) with an eigenfunction expansion. Determine D such that

$$U(.5, .5) = .5.$$

What do $U(.5, .5) = .5$ and the value of D imply about the relationship between u, x, t, and the parameters of problem i).

6.5) Determine a quadratic $P_2(x)$ such that the solution of

$$\mathcal{L}u = u_{xx} - u_t = 0$$

$$u(0, t) = \sin t, \qquad u(1, 0) = 0$$
$$u(x, 0) = 0$$

minimizes

$$\|u(x, 1) - P_2(x)\|_{L_2(0,1)}.$$

6.6) Solve in the subspace of the first N eigenfunctions the following problem:
Find $u_0(x)$ such that the solution of

$$\mathcal{L}u = u_{xx} - u_t = 0$$

$$u_x(0, t) = u(2, t) = 0$$
$$u(x, 0) = u_0(x)$$

satisfies

$$P_N u(x, 1) = P_N x^2 (2 - x).$$

Compute the solution for $N = 1, 2, 3$ and infer its behavior as N becomes large.

6.7) Combine numerical integration and differential equations methods, if necessary, to solve in the subspace of the first two eigenfunctions the reaction diffusion problems

i) $$\mathcal{L}u = u_{xx} - u_t = u^3$$

and

ii) $$\mathcal{L}u = u_{xx} - u_t = \frac{1}{1+u}$$

both subject to

$$u_x(0, t) = u_x(1, t) = 0$$
$$u(x, 0) = \cos \pi x.$$

Chapter 7

One-Dimensional Wave Equation

7.1 Applications of the eigenfunction expansion method

Following the format of Chapter 6 we examine the application of the eigenfunction expansion approach to the one-dimensional wave equation. Examples involve vibrating strings, chains, and pressure and electromagnetic waves. We comment on the convergence of the approximate solution to the analytic solution and conclude by showing that the equations to be solved in the eigenfunction expansion method are identical to those which arise when Duhamel's principle is combined with the traditional product ansatz of separation of variables.

Example 7.1 A vibrating string with initial displacement.

The vibrating string problem served in Chapter 5 to introduce eigenfunction expansions for the wave equation. Here we shall use this simple setting to examine the method in some detail.

Suppose that a string of length L with fixed ends is displaced from its equilibrium position and let go at time $t = 0$. We wish to study the subsequent motion of the string.

Newton's second law and a small amplitude assumption lead to the following mathematical model for the displacement $u(x, t)$ of the string at point x and time t from the equilibrium $u \equiv 0$:

$$\mathcal{L}u \equiv u_{xx} - \frac{1}{c^2} u_{tt} = 0 \tag{7.1}$$

$$u(0, t) = u(L, t) = 0$$

$$u(x,0) = u_0(x)$$

$$u_t(x,0) = 0$$

where the wave speed c is a known constant depending on the density of the string and the tension applied to it. For definiteness let us assume that

$$u_0(x) = \begin{cases} \frac{x}{A}, & 0 < x < A \\ \frac{L-x}{L-A}, & A < x < L \end{cases}$$

so that the string is "plucked" at the point A. With this very problem began the development of the method of separation of variables two hundred and fifty years ago [2]. The discussion of Chapter 5 applies immediately. The associated Sturm–Liouville problem is

$$\phi'' = \mu\phi$$

$$\phi(0) = \phi(L) = 0$$

and the approximating problem

$$\mathcal{L}u \equiv u_{xx} - \frac{1}{c^2} u_{tt} = 0 \tag{7.2}$$

$$u(0,t) = u(L,t) = 0$$

$$u(x,0) = P_N u_0(x)$$

$$u_t(x,0) = 0$$

has the solution

$$u_N(x,t) = \sum_{n=1}^{N} \hat{\alpha}_n \cos c\lambda_n t \phi_n(x) \tag{7.3}$$

where

$$\phi_n(x) = \sin \lambda_n x, \quad \lambda_n = \frac{n\pi}{L}$$

and

$$\hat{\alpha}_n = \frac{\langle u_0, \phi_n \rangle}{\langle \phi_n, \phi_n \rangle}.$$

We observe that each term of (7.3) corresponds to a standing wave with amplitude $\hat{\alpha}_n \phi_n(x)$ which oscillates with frequency $\frac{c\lambda_n}{2\pi}$. In other words, $u_N(x,t)$ is the superposition of standing waves.

Alternatively, we can use the identity

$$\cos c\lambda_n t \sin \lambda_n x = \frac{1}{2} \left[\sin \lambda_n (x + ct) + \sin \lambda_n (x - ct) \right]$$

and rewrite (7.3) in the form

$$u_N(x,t) = \frac{1}{2} \sum_{n=1}^{N} \hat{a}_n \left[\sin \lambda_n (x + ct) + \sin \lambda_n (x - ct) \right] \qquad (7.4)$$

so that u_N is also the superposition of traveling waves moving to the left and right with speed c.

We know from Section 4.4 that

$$P_N u_0(x) = \sum_{n=1}^{N} \hat{a}_n \phi_n(x)$$

is also the truncated Fourier series of the $2L$ periodic odd function $\hat{u}_0(x)$ which coincides with $u_0(x)$ on $[0, L]$. The boundary conditions $u_0(0) = u_0(L) = 0$ guarantee that $\hat{u}_0(x)$ is continuous and piecewise smooth so that

$$\hat{u}_0(x) = \lim_{N \to \infty} \sum_{n=1}^{N} \hat{a}_n \phi_n(x) \quad \text{on } (-\infty, \infty).$$

It follows now from (7.4) that

$$\lim_{N \to \infty} u_N(x,t) = \frac{1}{2} \left[\hat{u}_0(x + ct) + \hat{u}_0(x - ct) \right]. \qquad (7.5)$$

As an illustration we show in Fig. 7.1 the standing wave $\hat{a}_5 \cos c\lambda_5 t \phi_5(x)$ and the left-traveling wave $\frac{1}{2} \hat{a}_5 \sin \lambda_5 (x - ct)$ for $t = .4$ and $t = .8$, as well as the solution $u_{10}(x,t)$ for $t = 0, .4$, and $.8$ for the following data: $c = L = 1$ and

$$u_0(x) = \begin{cases} 3x, & x \in [0, 1/3] \\ \frac{3}{2}(1 - x), & x \in (1/3, 1] \end{cases}$$

$$u_1(x) = 0$$

so that

$$\hat{a}_n = \frac{9 \sin \frac{n\pi}{3}}{(n\pi)^2}.$$

For the plucked string we see that the $2L$ periodic odd extension of $u_0(x)$ has a discontinuous first derivative at $x = A + 2kL$ and $x = -A + 2kL$ for any integer k. Hence the d'Alembert solution for the plucked string is not a continuously differentiable function of time and space and hence not a classical solution. On the other hand, the projections of u_0 and u_1 into the span of finitely many eigenfunctions are infinitely differentiable so that $u_N(x,t)$ is infinitely differentiable. $P_N u_0$ describes a string with continuous curvature and may model the plucked string better than the mathematical abstraction u_0. Hence $u_N(x,t)$ is a meaningful solution.

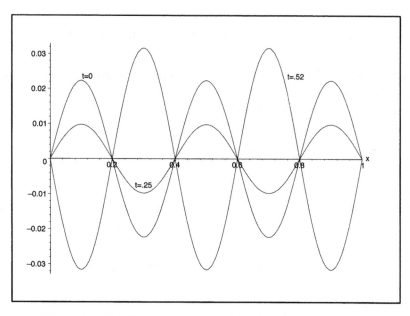

Figure 7.1: (a) Standing wave $\alpha_5(t)\phi_5(x)$ at $t = 0, .25, .52$.

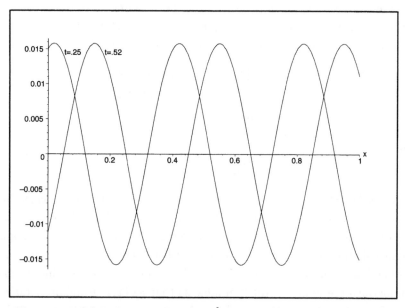

Figure 7.1: (b) Left-traveling wave $\frac{1}{2}\hat{\alpha}_5 \sin \lambda_5(x - t)$ at $t = .25, .52$.

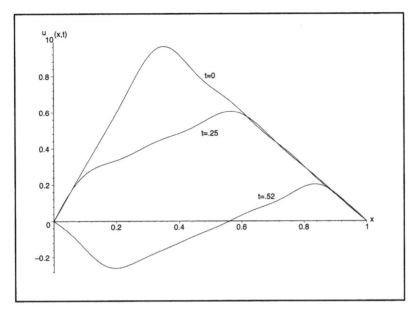

Figure 7.1: (c) Solution $u_{10}(x,t)$ for a plucked string at $t = 0, .25, .52$.

We note that such projections are always smoothing the data, even when the initial conditions are not consistent with the boundary conditions. For instance, if

$$u(0,t) = 0, \qquad t > 0$$

$$\lim_{x \to 0} u_0(x) \neq 0,$$

then there cannot be a classical solution with continuous $u(x,t)$ on $D = [0, L] \times [0, T]$. The eigenfunction method will apply and yield a smooth solution. $P_N u_0$ will, however, show a Gibbs phenomenon at $x = 0$. Unlike in the diffusion problems of Chapter 6 we have no smoothing of the solution of the wave equation with time so that the Gibbs phenomenon will persist and travel back and forth through the interval with time. In fact, if we choose

$$u_0(x) = 1 - x,$$

then (7.4) shows that one half of the Gibbs phenomenon initially at $x = 0$ will appear at a given point $x_0 \in (0,1)$ whenever

$$x_0 + t = 2k \qquad \text{for an integer } k,$$

$$x_0 - t = 2p \qquad \text{for an integer } p.$$

This behavior is illustrated in Fig. 7.1d where the d'Alembert solution $u(.5,t)$ and the eigenfunction expansion solution $u_{50}(.5,t)$ is plotted for $t \in (0,10)$ and $c = L = 1$.

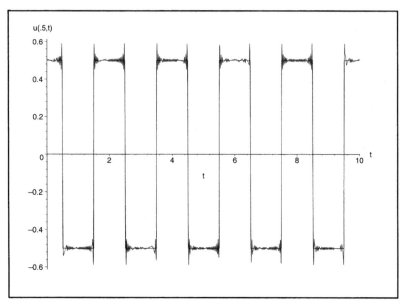

Figure 7.1: (d) Persistence of the Gibbs phenomenon due to inconsistent boundary and initial data. Shown are the d'Alembert solution (square wave) and the separation of variables solution at $x = .5$.

Example 7.2 A vibrating string with initial velocity.

We consider

$$\mathcal{L}u \equiv u_{xx} - \frac{1}{c^2} u_{tt} = 0$$

$$u(0, t) = u(L, t) = 0$$

$$u(x, 0) = 0$$

and with initial velocity

$$u_t(x, 0) = u_1(x)$$

for some smooth function $u_1(x)$. Since $u_t(0, t) = u_t(L, t) = 0$, we shall require that $u_1(0) = u_1(L) = 0$ in order to allow a smooth solution of the wave equation.

Eigenvalues and eigenfunctions are the same as in Example 7.1. If the initial condition is approximated by

$$u_t(x, 0) = P_N u_1(x) = \sum_{n=1}^{N} \hat{\beta}_n \phi_n(x),$$

then the corresponding approximating solution is

$$u_N(x, t) = \sum_{n=1}^{N} \alpha_n(t) \phi_n(x)$$

where

$$-\lambda_n^2 \alpha_n - \frac{1}{c^2}\alpha_n'' = 0$$

$$\alpha_n(0) = 0, \qquad \alpha_n'(0) = \hat{\beta}_n$$

so that

$$\alpha_n(t) = \frac{\hat{\beta}_n}{c\lambda_n} \sin c\lambda_n t.$$

We see from

$$u_N(x,t) = \sum_{n=1}^{N} \frac{\hat{\beta}_n}{c\lambda_n} \sin c\lambda_n t \sin \lambda_n x$$

that we again have a superposition of standing waves. If instead we use

$$\sin c\lambda_n t \sin \lambda_n x = \frac{1}{2}\left[\cos \lambda_n(x - ct) - \cos \lambda_n(x + ct)\right],$$

we obtain the superposition of traveling waves

$$u_N(x,t) = \frac{1}{2c}\sum_{n=1}^{N}\frac{\hat{\beta}_n}{\lambda_n}\left[\cos \lambda_n(x - ct) - \cos \lambda_n(x + ct)\right]. \qquad (7.6)$$

As in Example 7.1 we observe that

$$\lim_{N\to\infty}\sum_{n=1}^{N}\hat{\beta}_n \sin \lambda_n x = \hat{u}_1(x)$$

where $\hat{u}_1(x)$ is the $2L$ periodic odd function which coincides with $u_1(x)$ on $[0, L]$. Since the Fourier series can be integrated term by term, we find that

$$\int_{x-ct}^{x+ct}\hat{u}_1(s)ds = \lim_{N\to\infty}\sum_{n=1}^{N}\frac{\hat{\beta}_n}{\lambda_n}\left[\cos \lambda_n(x - ct) - \cos \lambda_n(x + ct)\right].$$

Thus (7.6) leads to

$$\lim_{N\to\infty} u_N(x,t) = \frac{1}{2c}\int_{x-ct}^{x+ct}\hat{u}_1(s)ds. \qquad (7.7)$$

The expressions (7.5) and (7.7) imply that the problem

$$\mathcal{L}u \equiv u_{xx} - \frac{1}{c^2}u_{tt} = 0$$

$$u(0,t) = u(L,t) = 0$$

$$u(x,0) = u_0(x)$$

$$u_t(x,0) = u_1(x)$$

has the solution

$$u(x,t) = \frac{1}{2}\left[u_0(x+ct) + u_0(x-ct) + \frac{1}{c}\int_{x-ct}^{x+ct} u_1(s)ds\right] \tag{7.8}$$

where u_0 and u_1 are identified with their $2L$ periodic extensions to $(-\infty,\infty)$.

The expression (7.8) is, of course, the d'Alembert solution which was derived in Section 1.5 without periodicity conditions on u_0 and u_1. Since the solution $u(x,t)$ of (7.8) depends on the data over $[x-ct, x+ct]$ only, and since L does not appear explicitly in (7.8), we can let $L \to \infty$ and conclude that (7.8) also holds for nonperiodic functions on $(-\infty,\infty)$ so that we have an alternate derivation of the d'Alembert solution of Chapter 1.

Example 7.3 A forced wave and resonance.

Suppose a uniform string of length L is held fixed at $x = 0$ and oscillated at $x = L$ according to

$$u(L,t) = A\sin\omega t.$$

We shall impose initial conditions

$$u(x,0) = 0, \qquad u_t(x,0) = A\omega\frac{x}{L}$$

which are consistent with the boundary data. We wish to find the motion of the string for $t > 0$.

In order to use an eigenfunction expansion we need homogeneous boundary conditions. If we set

$$v(x,t) = A\sin\omega t\left(\frac{x}{L}\right)$$

and

$$w(x,t) = u(x,t) - v(x,t),$$

then

$$\mathcal{L}w \equiv w_{xx} - w_{tt} = \mathcal{L}u - \mathcal{L}v = v_{tt} = -A\omega^2\sin\omega t\left(\frac{x}{L}\right)$$

where for ease of notation we have set $c = 1$. The boundary and initial conditions are

$$w(0,t) = w(L,t) = 0$$

$$w(x,0) = u(x,0) - v(x,0) = 0$$

$$w_t(x,0) = u_t(x,0) - v_t(x,0) = 0.$$

The associated eigenvalue problem is the same as in Example 7.1 and the approximating problem is

$$\mathcal{L}w \equiv w_{xx} - w_{tt} = \sin \omega t \sum_{n=1}^{N} \hat{\gamma}_n \phi_n(x)$$

where

$$\hat{\gamma}_n = -\frac{A}{L} \omega^2 \frac{\langle x, \phi_n \rangle}{\langle \phi_n, \phi_n \rangle} = \frac{2}{L} \frac{(-1)^n A \omega^2}{\lambda_n},$$

and

$$w(0, t) = w(L, t) = 0$$
$$w(x, 0) = 0$$
$$w_t(x, 0) = 0.$$

This problem has the solution

$$w_N(x, t) = \sum_{n=1}^{N} \alpha_n(t) \phi_n(x)$$

where

$$-\lambda_n^2 \alpha_n(t) - \alpha_n''(t) = \hat{\gamma}_n \sin \omega t$$
$$\alpha_n(0) = 0$$
$$\alpha_n'(0) = 0.$$

Then $\alpha_n(t)$ can be written in the form

$$\alpha_n(t) = c_1 \cos \lambda_n t + c_2 \sin \lambda_n t + \alpha_{np}(t)$$

where $\alpha_{np}(t)$ is a particular integral. Let us assume at this stage that $\omega \neq \lambda_n$ for any n. We can guess an $\alpha_{np}(t)$ of the form

$$\alpha_{np}(t) = C \sin \omega t.$$

If we substitute $\alpha_{np}(t)$ into the differential equation, we find that

$$C = \frac{\hat{\gamma}_n}{\omega^2 - \lambda_n^2}.$$

Determining c_1 and c_2 so that $\alpha_n(t)$ satisfies the initial conditions we obtain

$$\alpha_n(t) = \frac{\hat{\gamma}_n}{\lambda_n(\omega^2 - \lambda_n^2)} [\lambda_n \sin \omega t - \omega \sin \lambda_n t]$$

so that

$$u_N(x,t) = \sum_{n=1}^{N} \frac{\hat{\gamma}_n}{\lambda_n(\omega^2 - \lambda_n^2)} \left[\lambda_n \sin\omega t - \omega \sin\lambda_n t\right] \phi_n(x) + v(x,t).$$

Let us now suppose that the string is driven at a frequency ω which is close to λ_k for some k, say

$$\omega = \lambda_k + \epsilon, \qquad 0 < \epsilon \ll 1.$$

If we rewrite

$$\alpha_k(t) = \frac{\hat{\gamma}_k}{\lambda_k} \left[\frac{\lambda_k \sin\omega t - \omega \sin\lambda_k t}{\omega^2 - \lambda_k^2}\right]$$

as

$$\alpha_k(t) = \frac{\hat{\gamma}_k}{\lambda_k} \frac{\left[\omega(\sin\omega t - \sin\lambda_k t) + (\lambda_k - \omega)\sin\omega t\right]}{\omega^2 - \lambda_k^2}$$

and apply the identity

$$\sin x - \sin y = 2\sin\left(\frac{x-y}{2}\right)\cos\left(\frac{x+y}{2}\right),$$

we obtain

$$\alpha_k(t) = \frac{\hat{\gamma}_k\omega}{\lambda_k(\omega+\lambda_k)} \frac{2}{\epsilon}\sin\frac{\epsilon t}{2}\cos\left(\frac{\omega+\lambda_k}{2}\right)t - \frac{\hat{\gamma}_k \sin\omega t}{\lambda_k(\omega+\lambda_k)}.$$

The first term gives rise to a standing wave which oscillates with frequency $\frac{(\omega+\lambda_k)}{4\pi} \cong \frac{\lambda_k}{2\pi}$ but whose amplitude $\left[\frac{\hat{\gamma}_k\omega}{\lambda_k(\omega+\lambda_k)}\frac{2}{\epsilon}\sin\frac{\epsilon t}{2}\right]\phi_k(x)$ rises and falls slowly in time with frequency $\frac{\epsilon}{4\pi}$ which gives the motion a so-called "beat." Finally, we observe that if $\epsilon \to 0$, i.e., $\omega \to \lambda_k$, then l'Hospital's rule applied to the first term of $\alpha_k(t)$ yields

$$\alpha_k(t) = \frac{\hat{\gamma}_k}{2\lambda_k^2} \left[\lambda_k t \cos\lambda_k t - \sin\lambda_k t\right]$$

and after substituting for $\hat{\gamma}_k$

$$\alpha_k(t) = (-1)^k \frac{2A}{L} \left[\frac{t\cos\lambda_k t}{2} - \frac{\sin\lambda_k t}{2\lambda_k}\right]\phi_k(x).$$

Hence the amplitude of $\alpha_k(t)\phi_k(x)$ grows linearly with time and will eventually dominate all other terms in the eigenfunction expansion of $u_N(x,t)$. This phenomenon is called resonance and the string is said to be driven at the resonant frequency λ_k. An illustration of a beat is given in Fig. 7.3 where the amplitude $\alpha_k(t)$ of the standing wave $\alpha_k(t)\phi_k(x)$ is shown.

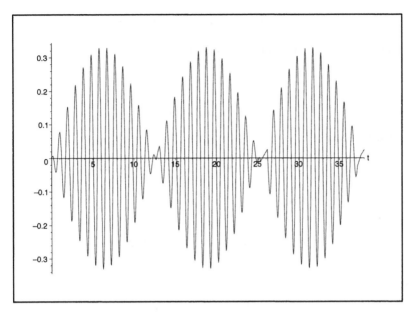

Figure 7.3: Amplitude of the ϵ-dependent part of $\alpha_2(t)$ near resonance. The amplitude waxes and wanes, producing a beat. $\omega = \lambda_2 + \epsilon$, $\epsilon = .5$, $\hat{\gamma}_2 = \phi_2 = 1$.

Example 7.4 Wave propagation in a resistive medium.

Suppose a string vibrates in a medium which resists the motion with a force proportional to the velocity and the displacement of the string. Newton's second law then leads to

$$\mathcal{L}u \equiv u_{xx} - Bu_{tt} - Cu_t - Du = 0$$

for the displacement $u(x,t)$ from the equilibrium position where B, C, and D are positive constants. (We remark that the same equation is also known as the telegraph equation and describes the voltage or current of an electric signal traveling along a lossy transmission line.) We shall impose the boundary and initial conditions of the last example

$$u(0,t) = 0, \qquad u(L,t) = A\sin\omega t$$

$$u(x,0) = 0, \qquad u_t(x,0) = \frac{A\omega}{L}x$$

and solve for the motion of the string.

If $v(x,t) = \frac{x}{L}A\sin\omega t$, then $w(x,t) = u(x,t) - v(x,t)$ satisfies

$$\mathcal{L}w = \mathcal{L}u - \mathcal{L}v = Bv_{tt} + Cv_t + Dv = \frac{Ax}{L}\left[(-B\omega^2 + D)\sin\omega t + C\omega\cos\omega t\right]$$

$$w(0,t) = w(L,t) = 0$$

$$w(x,0) = w_t(L,t) = 0.$$

The associated eigenvalue problem is again

$$\phi''(x) = \mu\phi(x)$$

$$\phi(0) = \phi(L) = 0.$$

The eigenfunctions are

$$\phi_n(x) = \sin\lambda_n x \quad \text{with} \quad \lambda_n = \frac{n\pi}{L}, \qquad \mu_n = -\lambda_n^2.$$

The approximating problem is

$$\mathcal{L}w = \sum_{n=1}^{N} \left[\hat{\gamma}_n \sin\omega t + \hat{\delta}_n \cos\omega t \right] \phi_n(x)$$

where

$$\hat{\gamma}_n = \frac{A(D - B\omega^2)}{L} \frac{\langle x, \phi_n \rangle}{\langle \phi_n, \phi_n \rangle}$$

$$\hat{\delta}_n = \frac{AC\omega}{L} \frac{\langle x, \phi_n \rangle}{\langle \phi_n, \phi_n \rangle}.$$

The approximating problem has the solution

$$w_N(x,t) = \sum_{n=1}^{N} \alpha_n(t)\phi_n(x)$$

where $\alpha_n(t)$ has to satisfy the initial value problem

$$-\lambda_n^2 \alpha_n(t) - B\alpha_n''(t) - C\alpha_n'(t) - D\alpha_n(t) = \hat{\gamma}_n \sin\omega t + \hat{\delta}_n \cos\omega t$$

$$\alpha_n(0) = 0, \qquad \alpha_n'(0) = 0.$$

It is straightforward to integrate this equation numerically. Moreover, as the following discussion shows, its solution for $C > 0$ is a decaying oscillatory function of time and hence can be computed quite accurately. A numerical integration would become necessary if the resisting forces in this model are nonlinear. Such forces would generally lead to a coupling of the initial value problems for $\{\alpha_n(t)\}$, but such nonlinear systems are solved routinely with explicit numerical methods.

It also is straightforward, and useful for analysis, to solve the equation analytically. Its solution has the form

$$\alpha_n(t) = c_{1n}\alpha_{n1}(t) + c_{2n}\alpha_{n2}(t) + \alpha_{np}(t)$$

where $\alpha_{n1}(t)$ and $\alpha_{n2}(t)$ are complementary solutions of the homogeneous equation and $\alpha_{np}(t)$ is any particular integral. For the complementary functions we try to fit an exponential function of the form

$$\alpha_c(t) = e^{rt}.$$

Substitution into the differential equation shows that r must be chosen such that

$$Br^2 + Cr + (D + \lambda_n^2) = 0.$$

This quadratic has the roots

$$r_{1,2} = \frac{-C \pm \sqrt{C^2 - 4B(D + \lambda_n^2)}}{2B}.$$

If $r_1 \neq r_2$, we may take

$$\alpha_{n1}(t) = e^{r_1 t}, \qquad \alpha_{n2}(t) = e^{r_2 t}.$$

The particular integral is best found with the method of undetermined coefficients. If we substitute

$$\alpha_{np}(t) = d_{1n} \sin \omega t + d_{2n} \cos \omega t$$

into the differential equation, then we require

$$+ B\omega^2 (d_{1n} \sin \omega t + d_{2n} \cos \omega t) - C\omega(d_{1n} \cos \omega t - d_{2n} \sin \omega t)$$
$$- (D + \lambda_n^2)(d_{1n} \sin \omega t + d_{2n} \cos \omega t) = \hat{\gamma}_n \sin \omega t + \hat{\delta}_n \cos \omega t.$$

Equating the coefficients of the trigonometric terms we find

$$\begin{pmatrix} B\omega^2 - (D + \lambda_n^2) & C\omega \\ -C\omega & B\omega^2 - (D + \lambda_n^2) \end{pmatrix} \begin{pmatrix} d_{1n} \\ d_{2n} \end{pmatrix} = \begin{pmatrix} \hat{\gamma}_n \\ \hat{\delta}_n \end{pmatrix}.$$

For $C > 0$ this equation has a unique solution so that d_{1n} and d_{2n} may be assumed known. Finally, we need to determine the coefficients c_{1n} and c_{2n} so that $\alpha_n(t)$ satisfies the initial conditions. We obtain the conditions

$$c_{1n} + c_{2n} = -d_{2n}$$

$$r_1 c_{1n} + r_2 c_{2n} = -\omega d_{1n}.$$

As long as $r_1 \neq r_2$ this system has the unique solution

$$c_{1n} = \frac{\omega d_{1n} - r_2 d_{2n}}{r_2 - r_1}$$

$$c_{2n} = \frac{r_1 d_{2n} - \omega d_{1n}}{r_2 - r_1}$$

so that

$$\alpha_n(t) = \frac{(\omega d_{1n} - r_2 d_{2n})e^{r_1 t} + (r_1 d_{2n} - \omega d_{1n})e^{r_2 t}}{r_2 - r_1} + d_{1n} \sin \omega t + d_{2n} \cos \omega t.$$

It is possible that for some index k the two roots r_1 and r_2 are the same. Then the first term for $\alpha_k(t)$ is indeterminate and must be evaluated with l'Hospital's rule. In analogy to mechanical systems we may say that the nth mode of our approximate solution is overdamped, critically damped, or underdamped if $C^2 - 4B(D + \lambda_n^2)$ is positive, zero, or negative, respectively. Using the trigonometric identities already employed in Example 7.1 (or complex exponentials) we see from $w_N(x,t) = \sum_{n=1}^{N} \alpha_n(t)\phi(x)$ that the term $\alpha_{np}(t)\phi_n(x)$ contributes traveling waves moving right and left with speed $\frac{\omega}{\lambda_n}$ and wave length $\frac{2\pi}{\lambda_n}$. For $C > 0$ the complementary solution in an overdamped or critically damped mode adds an exponentially decaying stationary solution of the form

$$e^{r_{1,2}t \pm i\lambda_n x}.$$

For sufficiently large n the nth mode will be underdamped because $\lambda_n \to \infty$ as $n \to \infty$. If we write

$$\alpha_n(t) = A_{1n}e^{r_1 t} + A_{2n}e^{r_2 t} + \alpha_{np}(t)$$

where

$$r_{1,2} = \frac{-C}{2B} \pm i \frac{\sqrt{4B(D + \lambda_n^2) - C^2}}{2B} = \frac{-C}{2B} \pm iE,$$

then the complementary part of $\alpha_n(t)\phi_n(x)$ contributes terms like

$$e^{\frac{-C}{2B}t \pm i(\lambda_n x \pm Et)}$$

which describes exponentially decaying traveling waves moving right and left with speed $\frac{E}{\lambda_n}$ and wave length $\frac{2\pi}{\lambda_n}$. We note that if we write

$$e^{-\frac{C}{2B}t + i(\lambda_n x - Et)}$$

formally as a plane wave

$$e^{i(kx - \delta t)},$$

then

$$k = \lambda_n$$

and

$$\delta = E - \frac{Ci}{2B}.$$

It follows that

$$\left(\delta + \frac{Ci}{2B}\right)^2 = E^2 = \frac{4B(D+k^2) - C^2}{4B^2}$$

or

$$\delta^2 + i\frac{C}{B}\delta - \frac{(D+k^2)}{B} = 0.$$

This equation is known as the dispersion relation for the telegraph equation. Conversely, any plane wave satisfying the dispersion relation is a solution of the unforced telegraph equation on the real line (for a discussion of the meaning and importance of dispersion see, e.g., [3]).

Example 7.5 Oscillations of a hanging chain.

This example shows that special functions like Bessel functions also arise for certain problems in cartesian coordinates. A chain of length L with uniform density is suspended from a (frictionless) hook and given an initial displacement and velocity which are assumed to lie in the same plane. We wish to study the subsequent motion of the chain.

Let $u(x,t)$ be the displacement from the equilibrium position where the coordinate x is measured from the free end of the chain at $x = 0$ vertically upward to the fixed end at $x = L$. Then the tension in the chain at x is proportional to the weight of the chain below x. After scaling, Newton's second law leads to the following mathematical model for the motion of the chain [4]:

$$\mathcal{L}u \equiv (xu_x)_x - u_{tt} = 0$$

$$u(L,t) = 0$$
$$u(x,0) = u_0(x)$$
$$u_t(x,0) = u_1(x).$$

The associated eigenvalue problem is

$$(x\phi'(x))' = \mu\phi(x)$$

$$\phi(L) = 0.$$

If we add the natural restriction that also $|\phi(0)| < \infty$, then we again have a singular Sturm–Liouville problem. Its solution is not obvious, but the arguments of Section 3.2 imply that all eigenvalues are necessarily nonpositive so that

$$\mu = -\lambda^2.$$

If we rewrite the eigenvalue problem in the form

$$x^2\phi''(x) + x\phi'(x) + \lambda^2 x\phi(x) = 0$$

$$|\phi(0)| < \infty, \qquad \phi(L) = 0$$

and match it against formula (3.10), we find immediately that it has the solutions

$$\left\{\mu_n, J_0\left(2\lambda_n\sqrt{x}\right)\right\}$$

where $\mu_n = -\lambda_n^2$ and where $\lambda_n 2\sqrt{L} = z_{n0}$. As before, z_{n0} is the nth zero of the zero order Bessel function $J_0(z)$. In summary, the eigenvalue problem associated with the hanging chain has the solution $\{\mu_n, \phi_n(x)\}$ with

$$\phi_n(x) = J_0(2\lambda_n\sqrt{x}), \qquad \lambda_n = \frac{z_{n0}}{2\sqrt{L}}, \qquad \mu_n = -\lambda_n^2.$$

We also know from our discussion of Example 6.10 that for $z = \sqrt{Lx}$

$$\int_0^L \phi_m(x)\phi_n(x)dx = \frac{2}{L}\int_0^L J_0\left(\frac{z_{n0}}{L}z\right) J_0\left(\frac{z_{m0}}{L}z\right) z\,dz = 0$$

so that the eigenfunctions $\{\phi_n(x)\}$ are orthogonal in $L_2(0, L)$. (We note that orthogonality with respect to the weight function $w(x) = 1$ is predicted by the Sturm–Liouville theorem of Chapter 3 if the eigenvalue problem were a regular problem given on an interval $[\epsilon, L]$ with $\epsilon > 0$.) The approximate solution is written as

$$u_N(x, t) = \sum_{n=1}^{N} \alpha_n(t)\phi_n(x).$$

Substitution into the wave equation leads to the initial value problems

$$-\lambda_n^2\phi_n(t) - \alpha_n''(t) = 0$$

$$\alpha_n(0) = \frac{\langle u_0(x), \phi_n(x)\rangle}{\langle \phi_n, \phi_n\rangle}, \qquad \alpha_n'(0) = \frac{\langle u_1(x), \phi_n(x)\rangle}{\langle \phi_n, \phi_n\rangle}$$

and the approximate solution

$$u_N(x, t) = \sum_{n=1}^{N} \left[\alpha_n(0)\cos\lambda_n t + \frac{\alpha_n'(0)}{\lambda_n}\sin\lambda_n t\right] J_0(2\lambda_n\sqrt{x}).$$

In Fig. 7.5 we show several positions of the chain as it swings from its initial straight-line displacement on the left to its maximum displacement on the right.

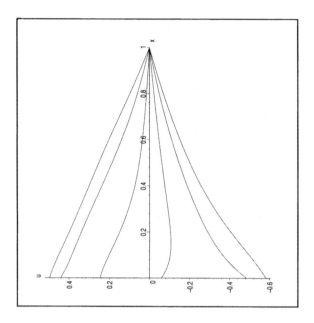

Figure 7.5: Position of the hanging chain as its end swings from its initial position to the right. The displacement $u(x,t)$ is shown for $t = 0, .5, 1, 1.5, 2, 2.4$; at $t = 2.4$, $u_t(0, 2.4) \approx 0$, $N = 4$, $L = 1$.

Example 7.6 Symmetric pressure wave in a sphere.

At time $t = 0$ a spherically symmetric pressure wave is created inside a rigid shell of radius R. We want to find the subsequent pressure distribution in the sphere.

Since there is no angular dependence, the mathematical model for the pressure $u(r,t)$ in the sphere is

$$u_{rr} + \frac{2}{r} u_r - \frac{1}{c^2} u_{tt} = 0.$$

We shall assume that the initial state can be described by

$$u(r,0) = u_0(r)$$

$$u_t(r,0) = u_1(r).$$

At all times we have to satisfy the symmetry condition

$$u_r(0,t) = 0.$$

Since the sphere is rigid, there is no pressure loss through the shell so we require

$$u_r(R,t) = 0.$$

The eigenvalue problem associated with this model is

$$\phi''(r) + \frac{2}{r}\phi'(r) = \mu\phi(r)$$

$$\phi'(0) = \phi'(R) = 0.$$

To ensure a finite pressure we shall require that $|\phi(0)| < \infty$. We encountered an almost identical problem in our discussion of heat flow in a sphere (Example 6.7) and already know that bounded solutions for nonzero λ have to have the form

$$\phi(r) = \frac{\sin\lambda r}{r}.$$

The boundary condition at $r = R$ requires that

$$\phi'(R) = \frac{R\lambda\cos\lambda R - \sin\lambda R}{2} = 0.$$

Hence the eigenvalues $u_n = -\lambda_n^2$ are determined from the roots of

$$f(\lambda) \equiv R\lambda\cos\lambda R - \sin\lambda R = 0.$$

The existence and distribution of roots of f were discussed in Example 6.6. We see by inspection that

$$f\left(\frac{n\pi}{R}\right)f\left(\frac{(n+1/2)\pi}{R}\right) < 0$$

so that there are countably many roots, and that

$$\lambda_n \to \left(\frac{\pi}{2} + n\pi\right)/R \qquad \text{as } n \to \infty$$

because the cosine term will dominate for large n. In addition we observe that for the boundary conditions of this application $\lambda_0 = \mu_0 = 0$ is an eigenvalue with eigenfunction $\phi_0(x) \equiv 1$.

It follows from the general theory that distinct eigenfunctions are orthogonal in $L_2(0, R, r^2)$. This orthogonality can also be established by simple integration. We see that for $\lambda_m \neq \lambda_n$

$$\langle\phi_m(r), \phi_n(r)\rangle = \int_0^R \frac{\sin\lambda_m r}{r}\frac{\sin\lambda_n r}{r}r^2 dr$$

$$= \frac{1}{\lambda_n^2 - \lambda_m^2}\left[\lambda_m\cos\lambda_m R\sin\lambda_n R - \lambda_n\cos\lambda_n R\sin\lambda_m R\right] = 0$$

in view of $f(\lambda_m) = f(\lambda_n) = 0$. It is straightforward to verify that this conclusion remains valid if $m = 0$ and $n \neq 0$. It follows that

$$u_N(r, t) = \sum_{n=1}^{N}\alpha_n(t)\phi_n(r)$$

where for $n \geq 0$

$$-(c\lambda_n)^2 \alpha_n(t) - \alpha_n''(t) = 0$$

$$\alpha_n(0) = \frac{\langle u_0(r), \phi_n \rangle}{\langle \phi_n, \phi_n \rangle}, \qquad \alpha_n'(0) = \frac{\langle u_1(r), \phi_n \rangle}{\langle \phi_n, \phi_n \rangle}.$$

Note that $\alpha_0(0)$ and $\alpha_0'(0)$ are the average values for the initial pressure and velocity over the sphere. The equations for $\alpha_n(t)$ are readily integrated. We obtain

$$u_N(x,t) = \alpha_0(0) + \alpha_0'(0)t + \sum_{n=1}^{N} \lambda_n \left[\alpha_n(0) \cos c\lambda_n t + \frac{\alpha_n'(0)}{c\lambda_n} \sin c\lambda_n t \right] \frac{\sin \lambda_n r}{\lambda_n r}.$$

For example, let us suppose that $R = 1$ and

$$u_0(r) = \begin{cases} \frac{\sin(10\pi r)}{r} & 0 < r < 1/10 \\ 0 & 1/10 \leq r < 1 \end{cases}$$

and

$$u_1(r) = 0.$$

Then

$$\alpha_0(0) = \frac{3}{100\pi}$$

$$\alpha_n(0) = 2 \frac{\frac{\sin \frac{(10\pi - \lambda_n)}{10}}{(10\pi - \lambda_n)} - \frac{\sin \frac{(10\pi + \lambda_n)}{10}}{(10\pi + \lambda_n)}}{2 - \sin(2\lambda_n)}$$

and

$$\alpha_n'(0) = 0.$$

Fig. 7.6 shows the pressure wave $u_{20}(r,t)$ at time $t = .5$ before it has reached the outer shell at $R = 1$, and after reflection at $t = 1.75$.

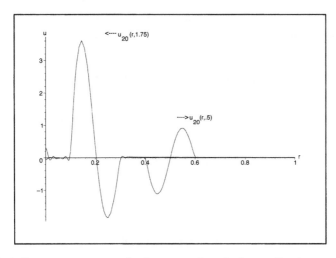

Figure 7.6: Pressure wave $u_{20}(r,t)$ at $t = .5$ and after reflection at $t = 1.75$.

Example 7.7 Controlling the shape of a wave.

The next example is reminiscent of constrained Hilbert space minimization problems discussed, for example, in [15]. The problem will be stated as follows.

Determine the "smallest" force $F(x,t)$ such that the wave $u(x,t)$ described by

$$u_{xx} - u_{tt} = F(x,t)$$

$$u(0,t) = u(L,t) = 0$$

$$u(x,0) = u_t(x,0) = 0$$

satisfies the final condition

$$u(x,T) = u_f(x)$$

$$u_t(x,T) = 0$$

where $u_f(x)$ is a given function and T is a given final time.

We shall ignore the deep mathematical questions of whether and in what sense this problem does indeed have a solution and concentrate instead on showing that we can actually solve the approximate problem formulated for functions of x which at any time belong to the subspace

$$M_N = \text{span}\{\phi_n(x)\}_{n=1}^N.$$

As before $\phi_n(x)$ is the nth eigenfunction of

$$\phi''(x) = \mu\phi(x)$$

$$\phi(0) = \phi(L) = 0,$$

i.e., $\phi_n(x) = \sin \lambda_n x$, $\lambda_n = \frac{n\pi}{L}$. To make the problem tractable we shall agree that the size of the force will be measured in the least squares sense

$$\|F\| = \left(\int_0^T \int_0^L F(x,t)^2 dx\, dt \right)^{1/2}.$$

The approximation to the above problem can then be formulated as:

Find

$$\hat{F}_N(x,t) = \sum_{n=1}^N \hat{\gamma}_n(t)\phi_n(x)$$

such that

$$\|\hat{F}_N\| \le \|F_N\|$$

for all $F_N(\cdot,t) \in M$ for which the solution u_N of

$$\mathcal{L}u = u_{xx} - u_{tt} = F_N(x,t)$$

$$u(0,t) = u(L,t) = 0$$

$$u(x,0) = u_t(x,0) = 0$$

satisfies

$$u(x,T) = P_N u_f(x)$$

$$u_t(x,T) = 0.$$

We know that F_N and u_N have the form

$$F_N(x,t) = \sum_{n=1}^{N} \gamma_n(t)\phi_n(x)$$

and

$$u_N(x,t) = \sum_{n=1}^{N} \alpha_n(t)\phi_n(x)$$

where

$$-\lambda_n^2 \alpha_n(t) - \alpha_n''(t) = \gamma_n(t)$$

$$\alpha_n(0) = \alpha_n'(0) = 0.$$

The variation of parameters solution for this problem can be verified to be

$$\alpha_n(t) = -\int_0^t \frac{1}{\lambda_n} \gamma_n(s) \sin \lambda_n(t-s) ds.$$

$u_N(x,T)$ will satisfy the final condition if $\gamma_n(t)$ is chosen such that

$$\alpha_n(T) = \frac{\langle u_f(x), \phi_n \rangle}{\langle \phi_n, \phi_n \rangle}$$

$$\alpha_n'(T) = 0.$$

Hence $\gamma_n(t)$ must be found such that

$$\int_0^T \gamma_n(s) \sin \lambda_n(T-s) ds = -\lambda_n \frac{\langle u_f, \phi_n \rangle}{\langle \phi_n, \phi_n \rangle} \qquad (7.9)$$

$$\int_0^T \gamma_n(s) \cos \lambda_n(T-s) ds = 0. \qquad (7.10)$$

Finally, we observe that

$$\|F_N\|^2 = \frac{L}{2} \sum_{n=1}^{N} \int_0^T \gamma_n(t)^2 dt$$

so that $\|F_N\|$ will be minimized whenever $\int_0^T \gamma_n(t)^2 dt$ is minimized for each n. But it is known from Theorem 2.13 that the minimum norm solution in $L_2[0, T]$ of the two constraint equations (7.9), (7.10) must be of the form

$$\hat{\gamma}_n(t) = c_{1n} \sin \lambda_n(T - t) + c_{2n} \cos \lambda_n(T - t).$$

Substitution into (7.9), (7.10) and integration with respect to t show that c_{1n} and c_{2n} must satisfy

$$\begin{pmatrix} \frac{T}{2} - \frac{\sin 2\lambda_n T}{4\lambda_n} & \frac{\sin^2 \lambda_n T}{2\lambda_n} \\ \frac{\sin^2 \lambda_n T}{2\lambda_n} & \frac{T}{2} + \frac{\sin 2\lambda_n T}{4\lambda_n} \end{pmatrix} \begin{pmatrix} c_{1n} \\ c_{2n} \end{pmatrix} = \begin{pmatrix} -\lambda_n \frac{\langle u_f, \phi_n \rangle}{\langle \phi_n, \phi_n \rangle} \\ 0 \end{pmatrix}.$$

We observe that the determinant of the coefficient matrix is

$$\frac{1}{4\lambda_n^2} \left[(T\lambda_n)^2 - \sin^2 \lambda_n T \right]$$

and hence never zero for $T > 0$. Thus each $\hat{\gamma}_n(t)$ is uniquely defined and the approximating problem is solved. Whether

$$F_N(x, t) = \sum_{n=1}^N \hat{\gamma}_n(t) \phi_n(x)$$

remains meaningful as $N \to \infty$ depends strongly on $u_f(x)$. It can be shown by actually solving the linear system for c_{1n} and c_{2n} that

$$c_{1n} \sim \frac{2\lambda_n \langle u_f, \phi_n \rangle}{T} \quad \text{and} \quad |c_{2n}| \ll |c_{1n}| \qquad \text{as } n \to \infty.$$

Integrability of u_f' and the consistency condition $u_f(0) = u_f(L) = 0$ yield $\lambda_n \langle u_f, \phi_n \rangle = \langle u_f', \cos \lambda_n x \rangle$, and Bessel's inequality guarantees that

$$\sum_{n=1}^N c_{1n}^2 \sim \sum_{n=1}^N \langle u_f', \cos \lambda_n x \rangle^2 \sim \|u_f'\|^2$$

so that F_N will converge in the mean square sense as $N \to \infty$.

Example 7.8 The natural frequencies of a uniform beam.

This example is chosen to illustrate that the eigenfunction approach depends only on the solvability of the eigenvalue problem, not on the order of the differential operators.

A mathematical model for the displacement $u(x, t)$ of a beam is [10]

$$\frac{\partial^4 u}{\partial x^4} + \frac{1}{c^2} \frac{\partial^2 u}{\partial t^2} = 0$$

$$u(0,t) = u_x(0,t) = 0$$

$$u_{xx}(L,t) = u_{xxx}(L,t) = 0.$$

The boundary conditions indicate that the beam is fixed at $x = 0$ and free at $x = L$. We want to find the natural frequencies of the beam.

The associated eigenvalue problem is

$$\phi^{(iv)}(x) = \mu\phi(x)$$

$$\phi(0) = \phi'(0) = 0$$

$$\phi''(L) = \phi'''(L) = 0.$$

It is straightforward to show as in Chapter 3 that the eigenvalue must be positive. For notational convenience we shall write

$$\mu = \lambda^4$$

for some positive λ. We observe that the function

$$\phi(x) = e^{rx}$$

will solve the differential equation if

$$r^4 = \lambda^4.$$

The four roots of the positive number λ^4 are

$$r_1 = \lambda, \qquad r_2 = -\lambda, \qquad r_3 = i\lambda, \quad \text{and} \quad r_4 = -i\lambda.$$

A general solution of the equation is then

$$\phi(x) = c_1 \sinh \lambda x + c_2 \cosh \lambda x + c_3 \sin \lambda x + c_4 \cos \lambda x.$$

Substitution into the boundary conditions leads to the system

$$\begin{pmatrix} 0 & 1 & 0 & 1 \\ 1 & 0 & 1 & 0 \\ \sinh \lambda L & \cosh \lambda L & -\sin \lambda L & -\cos \lambda L \\ \cosh \lambda L & \sinh \lambda L & -\cos \lambda L & \sin \lambda L \end{pmatrix} \begin{pmatrix} c_1 \\ c_2 \\ c_3 \\ c_4 \end{pmatrix} = 0.$$

It is straightforward to verify that the determinant of the coefficient matrix is zero if and only if

$$\cosh \lambda L \cos \lambda L + 1 = 0.$$

If we rewrite this equation as

$$f(x) \equiv \cos x + \frac{1}{\cosh x} = 0,$$

then f behaves essentially like $\cos x$ and has two roots in every interval $(n\pi - \pi/2, n\pi + \pi/2)$ for $n = 1, 3, 5, \ldots$. The first five numerical roots of $f(x) = 0$ are tabulated below.

i	x_i
1	1.87510
2	4.69409
3	7.85476
4	$10.9955 \cong 3\pi - \pi/2$
5	$14.1372 \cong 3\pi + \pi/2$

Since $\cosh x$ grows exponentially, all subsequent roots are numerically the roots of $\cos x$. Thus the eigenvalues for the vibrating beam are

$$\mu_n = \lambda_n^4 \qquad \text{where } \lambda_n = x_n/L.$$

The corresponding eigenfunctions satisfy $c_1 = -c_3$, $c_2 = -c_4$. A nontrivial solution for the coefficients is

$$(c_1, c_2, c_3, c_4) = (\cosh \lambda_n L + \cos \lambda_n L, -\sinh \lambda_n L - \sin \lambda_n L, -c_1, -c_2)$$

so that

$$\phi_n(x) = [\cosh \lambda_n L + \cos \lambda_n L][\sinh \lambda_n x - \sin \lambda_n x]$$
$$- [\sinh \lambda_n L + \sin \lambda_n L][\cosh \lambda_n x - \cos \lambda_n x]. \qquad (7.11)$$

An oscillatory solution of the beam equation is obtained when we write

$$u_n(x, t) = \alpha_n(t)\phi_n(x)$$

and compute $\alpha_n(t)$ such that

$$\lambda_n^4 \alpha_n(t) + \frac{1}{c^2} \alpha_n''(t) = 0.$$

It follows that

$$\alpha_n(t) = A_n \cos(c\lambda_n^2 t + B_n)$$

where the amplitude A_n and the phase B_n are the constants of integration. Each $u_n(x, t)$ describes a standing wave oscillating with the frequency $f_n = \frac{c\lambda_n^2}{2\pi}$.

The motion of a vibrating beam subject to initial conditions and a forcing function is found in the usual way by projecting the data into the span of the eigenfunctions and solving the approximating problem in terms of an eigenfunction expansion.

Example 7.9 A system of wave equations.

To illustrate how automatic the derivation and solution of the approximating problem has become we shall consider the following model problem:

$$\mathcal{L}_1 u = u_{xx} - c_1 u_{tt} = \chi(x)(u - v), \qquad x \in (0, 1)$$

$$\mathcal{L}_2 v = v_{xx} - c_2 v_{tt} = -\chi(x)(u - v), \qquad x \in (0, 2)$$

$$u(0, t) = v(0, t) = 0$$

$$u(1, t) = v_x(2, t) = 0$$

plus initial conditions like

$$\begin{aligned} u(x, 0) &= u_0(x) \\ u_t(x, 0) &= 0 \end{aligned} \qquad x \in (0, 1),$$

$$\begin{aligned} v(x, 0) &= v_0(x) \\ v_t(x, 0) &= 0 \end{aligned} \qquad x \in (0, 2).$$

Here $\chi(x)$ denotes the characteristic function of some interval I contained in $[0, 1]$, i.e.

$$\chi(x) = \begin{cases} 1 & x \in I \\ 0 & x \notin I. \end{cases}$$

The eigenvalue problem associated with $\mathcal{L}_1$ and the boundary conditions for u is

$$\phi'' = \mu \phi$$

$$\phi(0) = \phi(1) = 0$$

with solution

$$\phi_m(x) = \sin \lambda_m x, \qquad \lambda_m = m\pi, \qquad \mu_m = -\lambda_m^2, \qquad m = 1, \ldots, M.$$

The eigenvalue problem for $\mathcal{L}_2$ is

$$\psi'' = \mu \psi$$

$$\psi(0) = \psi'(2) = 0$$

and has the solution

$$\psi_n(x) = \sin \eta_n x, \qquad \eta_n = \frac{\frac{\pi}{2} + n\pi}{2}, \qquad \mu_n = -\eta_n^2, \qquad n = 0, \ldots, N.$$

The approximating problem is found by projecting the source term and the initial conditions for u into span$\{\phi_m\}$, and the corresponding terms for v into span$\{\psi_n\}$. We obtain

$$\mathcal{L}_1 u = P_M\{\chi(x)(u(x) - v(x))\}$$

$$u(x,0) = P_M u_0(x), \qquad u_t(x,0) = 0$$

$$\mathcal{L}_2 v = P_N\{-\chi(x)(u(x) - v(x))\}$$

$$v(x,0) = P_N v(x), \qquad v_t(x,0) = 0.$$

The approximating problem admits a solution $u_M \in \operatorname{span}\{\phi_m\}$ and $v_N \in \operatorname{span}\{\psi_n\}$ because, if we write

$$u_M(x) = \sum_{m=1}^{M} \alpha_m(t)\phi_m(x)$$

and

$$v_N(x) = \sum_{n=0}^{N} \beta_n(t)\psi_n(x)$$

and substitute these representations into the approximating problem, then we find that $\alpha_m(t)$ and $\beta_n(t)$ must satisfy the initial value problems

$$-\lambda_m^2 \alpha_m(t) - c_1 \alpha_m''(t) = \frac{\langle \chi(x)(u_M(x) - v_N(x)), \phi_m \rangle}{\langle \phi_m, \phi_m \rangle} \qquad (7.12)$$

$$= \sum_{i=1}^{M} C_{mi}^1 \alpha_i(t) + \sum_{j=0}^{N} D_{mj}^1 \beta_j(t)$$

$$-\eta_n^2 \beta_n(t) - c_2 \beta_n''(t) = \frac{\langle\langle -\chi(x)(u_M(x) - v_N(x)), \psi_m \rangle\rangle}{\langle\langle \psi_n, \psi_n \rangle\rangle}$$

$$= \sum_{i=1}^{M} C_{ni}^2 \alpha_i(t) + \sum_{j=0}^{N} D_{nj}^2 \beta_j(t)$$

$$\alpha_m(0) = \frac{\langle u_0, \phi_m \rangle}{\langle \phi_m, \phi_m \rangle}, \qquad \alpha_m'(0) = 0$$

$$\beta_n(0) = \frac{\langle\langle v_0, \psi_n \rangle\rangle}{\langle\langle \psi_n, \psi_n \rangle\rangle}, \qquad \beta_n'(0) = 0$$

where $\langle \ , \ \rangle$ and $\langle\langle \ , \ \rangle\rangle$ denote the inner products of $\mathcal{L}_2(0,1)$ and $\mathcal{L}_2(0,2)$, respectively. Specifically

$$C_{mi}^1 = 2 \int_I \phi_i(x)\phi_m(x)dx$$

$$D_{mj}^1 = -2 \int_I \psi_j(x)\phi_m(x)dx$$

$$C_{ni}^2 = -\int_I \phi_i(x)\psi_n(x)dx$$

$$D_{nj}^2 = \int_I \psi_j(x)\psi_n(x)dx.$$

Problem (7.12) is an initial value problem for a system of linear ordinary differential equations and has a unique solution $\{\alpha_m(t)\}$, $\{\beta_n(t)\}$. Thus, the approximating problem is solved (in principle).

We shall include the results of a preliminary numerical simulation with the data

$$c_1 = c_2 = 1, \qquad I = [.45, .55]$$

$$u_0(x) = \sin \lambda_1 x, \qquad v_0(x) = 0.$$

Shown in Figures 7.9 are the graphs of $u_3(.3, t)$ and $v_3(.3, t)$ for $t \in [0, 10]$. The pictures indicate that $u_3(.3, t)$ experiences a phase shift with time while $v_3(.3, t)$ shows no particular pattern. The results remain stable for long time runs. We have no information on u_N and v_N for $N > 4$ but note that the results for $N = 3$ and $N = 4$ are very close.

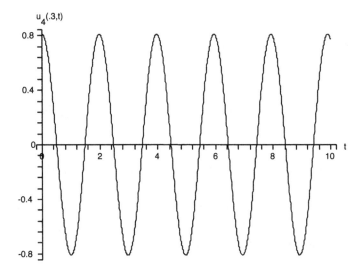

Figure 7.9: (a) Plot of $u_4(.3, t)$ vs. time. The graph shows a slow phase shift due to the coupling with $v(x, t)$, $I = [.45, .55]$.

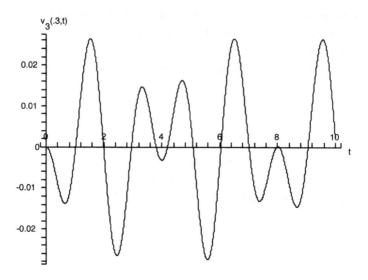

Figure 7.9: (b) Plot of $v_3(.3, t)$. The excitation of v is due solely to the coupling with u.

7.2 Convergence of $u_N(x, t)$ to the analytic solution

The convergence of the approximate solution u_N to the analytic solution u of the original problem can be examined with the help of the energy method which was introduced in Section 1.5 to establish uniqueness of the initial value problem for the wave equation. We shall repeat some of the earlier arguments here for the simple case of the one-dimensional wave equation. We consider the model problem

$$w_{xx} - w_{tt} = F(x, t)$$
$$w(0, t) = w(L, t) = 0$$
$$w(x, 0) = w_0(x)$$
$$w_t(x, t) = w_1(x). \tag{7.13}$$

Theorem 7.10 *Assume that the data of problem (7.13) are sufficiently smooth so that it has a smooth solution $w(x, t)$ on $D = \{(x, t) : 0 < x < L, 0 < t < T\}$*

for some $T > 0$. Then for $t \leq T$

$$\int_0^L [w_x^2(x,t) + w_t^2(x,t)]dx \leq e^t \int_0^L [w_1^2(x)$$

$$+ w_0'^2(x)]dx + \int_0^t \int_0^L e^{t-s} F^2(x,t)dx\,dt.$$

Proof. We multiply the wave equation by w_t

$$w_t w_{xx} - w_t w_{tt} = w_t F(x,t)$$

and use the smoothness of w to rewrite this expression in the form

$$(w_t w_x)_x - (w_{xt} w_x) - w_t w_{tt} = w_t F(x,t). \tag{7.14}$$

Since $w(0,t) = w(L,t) = 0$, it follows that $w_t(0,t) = w_t(L,t) = 0$. The integral of (7.14) with respect to x can then be written in the form

$$\frac{1}{2} \frac{d}{dt} \int_0^L (w_t^2 + w_x^2)dx = -\int_0^L w_t F(x,t)dx. \tag{7.15}$$

Using the algebraic-geometric inequality $2w_t F(x,t) \leq w_t^2 + F^2(x,t)$ and defining the "energy" integral

$$E(t) = \int_0^L [w_t^2(x,t) + w_x^2(x,t)]dx$$

we obtain from (7.15) the inequality

$$\frac{d}{dt} E(t) \leq E(t) + \|F(\cdot,t)\|^2 \tag{7.16}$$

where $\|F(\cdot,t)\|^2 = \int_0^L F^2(x,t)dx$. Gronwall's inequality applies to (7.16) and yields

$$E(t) \leq E(0)e^t + \int_0^t e^{t-s} \|F(\cdot,s)\|^2 ds$$

which was to be shown.

We note from (7.15) that if $F = 0$, then $\frac{dE}{dt}(t) = 0$ for all t and hence

$$E(t) = E(0).$$

We also observe that Schwarz's inequality (Theorem 2.4) allows the pointwise estimate

$$|w(x,t)| = \left| \int_0^x w_x(r,t)dr \right| \leq \sqrt{x}\,\|w_x(\cdot,t)\| < \sqrt{x}\sqrt{E(t)}$$

and that the Poincaré inequality of Example 4.20 yields

$$\|w(\cdot, t)\| < \frac{L}{\pi} \|w_x(\cdot, t)\| < \frac{L}{\pi} \sqrt{E(t)}.$$

These estimates are immediately applicable to the error

$$e_N(x, t) = w(x, t) - w_N(x, t)$$

where w solves (7.13) and w_N is the computed approximation obtained by projecting $F(\cdot, t)$, u_0, and u_1 into the span$\{\sin \lambda_n x\}_{n=1}^N$ with $\lambda_n = \frac{n\pi}{L}$.

Since $e_N(x, t)$ satisfies (7.13) with the substitutions

$$F \leftarrow F - P_N F, \quad w_0 \leftarrow w_0 - P_N w_0, \quad w_1 \leftarrow w_1 - P_N w_1,$$

we see from Theorem 7.10 that

$$\|e_x(\cdot, t)\|^2 + \|e_t(\cdot, t)\|^2 \le e^t \left[\|w_1 - P_N w_1\|^2 + \|(w_0 - P_N w_0)'\|^2 \right]$$
$$+ \int_0^t \int_0^L e^{t-s} \|F(\cdot, s) - P_N F(\cdot, s)\|^2 ds.$$

This estimate implies that if $P_N F(\cdot, t)$ converges in the mean square sense to $F(\cdot, t)$ uniformly with respect to t and $P_N w_0' \to w_0'$, then the computed solution converges pointwise and in the mean square sense to the true solution. But in contrast to the diffusion setting the error will not decay with time. If it should happen that $F \in \text{span}\{\sin \lambda_n x\}$ for all t, then the energy of the error remains constant and equal to the initial energy. If the source term $F - P_N F$ does not vanish, then the energy could conceivably grow exponentially with time.

7.3 Eigenfunction expansions and Duhamel's principle

The influence of $v(x, t)$ chosen to zero out nonhomogeneous boundary conditions imposed on the wave equation can be analyzed as in the case of the diffusion equation and will not be studied here. However, it may be instructive to show that the eigenfunction approach leads to the same equations as Duhamel's principle for the wave equation with time-dependent data so that again we only provide an alternative view but not a different computational method.

Consider the problem

$$w_{xx} - w_{tt} = F(x, t)$$
$$w(0, t) = w(L, t) = 0$$
$$w(x, 0) = 0$$
$$w_t(x, 0) = 0. \tag{7.17}$$

Duhamel's principle of Section 1.5 yields the solution

$$w(x, t) = \int_0^t W(x, t, s) ds$$

where

$$W_{xx}(x, t, s) - W_{tt}(x, t, s) = 0$$
$$W(0, t, s) = W(L, t, s) = 0$$
$$W(x, s, s) = 0$$
$$W_t(x, s, s) = -F(x, s).$$

The absence of a source term makes the calculation of a separation of variables solution of $W(x, t, s)$ straightforward.

An eigenfunction solution obtained directly from (7.17) is found in the usual way in the form

$$w_N(x, t) = \sum_{n=1}^N \alpha_n(t) \phi_n(x)$$

where $\alpha_n(t)$ solves the initial value problem

$$-\lambda_n^2 \alpha_n(t) - \alpha_n''(t) = \gamma_n(t)$$

$$\alpha_n(0) = \alpha_n'(0) = 0$$

with

$$\gamma_n(t) = \frac{\langle F(x, t), \phi_n \rangle}{\langle \phi_n, \phi_n \rangle}.$$

The variation of parameters solution for this problem is

$$\alpha_n(t) = -\frac{1}{\lambda_n} \int_0^t \sin \lambda_n(t - s) \gamma_n(s) ds.$$

If we set

$$W_n(x, t, s) = -\frac{1}{\lambda_n} \sin \lambda_n(t - s) \gamma_n(s) \phi_n(x),$$

then the eigenfunction solution is given by

$$w_N(x, t) = \sum_{n=1}^N \alpha_n(t) \phi_n(x) = \int_0^t \sum_{n=1}^N W_n(x, t, s).$$

By inspection

$$W_{n_{xx}} - W_{n_{tt}} = 0$$

$$W_n(x, s, s) = 0$$

$$W_{n_t}(x, s, s) = -\gamma_n(s)\phi_n(x) = -\frac{\langle F(x, s), \phi_n \rangle}{\langle \phi_n, \phi_n \rangle} \phi_n(x)$$

so that

$$\sum_{n=1}^{N} W_n(x, s, s) = -P_n F(x, s).$$

Hence both methods yield the same solution and require the evaluation of identical integrals.

Exercises

7.1) The air pressure above ambient in an organ pipe is modeled with

$$\mathcal{L}p \equiv p_{xx} - \frac{1}{c^2} p_{tt} = 0, \qquad 0 < x < L,$$

where c is the speed of sound. We suppose the pipe is closed at the bottom

$$p_x(0, t) = 0$$

and

i) Closed at the top so that $p_x(L, t) = 0$.

ii) Open at the top so that $p(L, t) = 0$.

For each case determine L so that the dominant frequency of the sound from the pipe is 440 Hertz. Use c at 20 C°.

7.2) Solve

$$\mathcal{L}u \equiv u_{xx} - \frac{1}{4}u_{tt} = 0, \qquad 0 < x < 4, \quad t > 0$$

$$u(0, t) = u(4, t) = 0$$

$$u(x, 0) = \max\{x(1 - x), 0\}$$

$$u_t(x, 0) = 0.$$

i) Compute the approximate solution $u_{10}(x, t)$.

ii) Find T_0 such that

$$\frac{\partial}{\partial x} u_{10}(4, t) \equiv 0 \quad \text{for } t \in [0, T_0].$$

iii) Find T_1 such the d'Alembert solution $u(x,t)$ of this problem satisfies

$$\frac{\partial}{\partial x} u(4,t) \equiv 0 \quad \text{for } t \in [0, T_1].$$

Why are both times not the same?

7.3) Consider

$$\mathcal{L}u \equiv u_{xx} - u_{tt} = 0$$

$$u(0,t) = \cos \omega t, \qquad u(1,t) = \cos \delta t$$

$$u(x,0) = 1$$

$$u_t(x,0) = 0.$$

Prove or disprove: for $\omega \neq \delta$ there is no resonance.

7.4) Consider

$$\mathcal{L}u \equiv u_{xx} - u_{tt} - Cu_t = 0$$

$$u(0,t) = \cos \omega t, \qquad u_x(1,t) = 0$$

$$u(x,0) = 1, \qquad u_t(x,0) = 0.$$

Find $u_N(1,t)$ and examine whether it shows a phase shift relative to the input $u(0,t)$.

7.5) For the hanging chain of Example 7.5 determine

i) $u(x,t^*)$ and $u_t(x,t^*)$ for $0 < x < 1$ for the first time t^* when $u(0,t) = 0$ has swung all the way to the right.

ii) $u(x,t^{**})$ and $u_t(x,t^{**})$ for $0 < x < 1$ for the second time t^{**} when $u(0,t) = 0$ has swung all the way to the right.

iii) Do the calculations show that $u(x,t)$ becomes periodic in time and space, either quickly or eventually?

7.6) Combine numerical integration with an explicit numerical routine for initial value problems for ordinary differential equations to give an eigenfunction solution of

$$\mathcal{L}u \equiv u_{xx} - u_{tt} = -\sin u$$

$$u(0,t) = u(1,t) = 0$$

$$u(x,0) = \sin \pi x$$

$$u_t(x,0) = 0.$$

(For a feasibility study you might choose the trapezoidal rule for the numerical integration and an explicit Euler method for finding $\alpha_n(t)$ for $n = 1, 2$.)

7.7) Solve the vibrating beam problem

$$\mathcal{L}u \equiv u_{xxxx} + u_{tt} = 0, \qquad 0 < x < 1, \quad t > 0$$

with one fixed end

$$u(0,t) = u_x(0,t) = 0$$

and

i) A free end

$$u_{xx}(1,t) = u_{xxx}(1,t) = 0$$

ii) An elastically supported end

$$u_{xxx}(1,t) + u(1,t) = 0, \qquad u_{xx}(1,t) = 0$$

and initial conditions

$$u(x,0) = -x$$
$$u_t(x,0) = 0.$$

Chapter 8

Potential Problems in the Plane

In principle, little will change when eigenfunction expansions are applied to potential problems. In practice, however, the mechanics of solving such problems tend to be more complicated because the expansion coefficients are found from boundary value problems which are more difficult to solve than the initial value problems arising in diffusion and wave propagation. In addition, a new issue of preconditioning arises from the strong coupling of the boundary data along each coordinate direction.

8.1 Applications of the eigenfunction expansion method

Example 8.1 The Dirichlet problem for the Laplacian on a rectangle.

Applications in steady-state heat transfer and electrostatics (among others) lead to the Dirichlet problem for Poisson's equation discussed in Section 1.3

$$\Delta u \equiv u_{xx} + u_{yy} = F(x, y), \qquad (x, y) \in D \qquad (8.1)$$

$$u(x, y) = g(x, y), \qquad (x, y) \in \partial D$$

where D is the rectangle $0 < x < a$, $0 < y < b$ and ∂D is its boundary.

If the boundary value g is a continuous function as we travel around ∂D and if F is continuous in D, then as stated in Section 1.3 the problem (8.1) has a unique solution which is continuous on $\bar{D}$ and twice continuously differentiable in D, i.e., a classical solution. However, as we shall see in Example 8.4, discontinuous boundary data do arise in applications which require a generalized solution.

195

If the boundary function g is twice continuously differentiable on opposite sides of the rectangle D, then the usual solution recipe of Chapter 5 can be followed. We zero out the boundary conditions, say at $x = 0$ and $x = a$, with the function

$$v(x,y) = g(x,y)$$

if g is defined and smooth on D, or

$$v(x,y) = g(0,y)\frac{a-x}{a} + g(a,y)\frac{x}{a}$$

if g is given on ∂D only, and solve the problem for

$$w(x,y) = u(x,y) - v(x,y)$$

in terms of the eigenfunctions $\{\sin \lambda_n x\}$ as described in Chapter 5.

This is the preferred approach in this text, but if the boundary function is, for example,

$$g(x,y) = xy|x - a/2||y - b/3|,$$

then v would not be differentiable and the problem cannot be recast into a Dirichlet problem for w.

For the potential equation there is an alternate way to zero out the boundary data on opposing sides of ∂D which does not demand smoothness of g and which does not affect the source term F in (8.1). Problem (8.1) can be solved "in principle" by splitting it into two problems. We write

$$u = u_1 + u_2$$

where

$$\Delta u_1 = F_1(x,y) \tag{8.2}$$
$$u_1(0,y) = u_1(a,y) = 0$$
$$u_1(x,0) = g(x,0), \quad u_1(x,b) = g(x,b),$$

and

$$\Delta u_2 = F_2(x,y)$$

$$u_2(0,y) = g(0,y), \qquad u_2(a,y) = g(a,y), \tag{8.3}$$
$$u_2(x,0) = u_2(x,b) = 0,$$

and

$$F_1(x,y) + F_2(x,y) = F(x,y).$$

We find an approximation u_{1M} by projecting $F_1(x, y)$, $g(x, 0)$, and $g(x, b)$ into the span$\{\phi_m(x)\}$ of the eigenfunctions of

$$\phi''(x) = \mu\phi(x)$$
$$\phi(0) = \phi(a) = 0$$

associated with (8.2), and an approximation u_{2N} by projecting $F_2(x, y)$, $g(0, y)$, and $g(a, y)$ into span$\{\psi_n(y)\}$ of the eigenfunctions of

$$\psi''(y) = \mu\psi(y)$$
$$\psi(0) = \psi(b) = 0$$

associated with (8.3), and solving for u_{1M} and u_{2N} as outlined in Chapter 5. The choice of F_1 and F_2 depends on how easily and accurately their projections can be computed. $u_{1M} + u_{2N}$ then is an approximation to the solution of (8.1). We shall refer to (8.2) and (8.3) as a "formal splitting" of the Dirichlet problem (8.1).

This formal splitting is routinely applied in texts on separation of variables and is justified because the solutions are thought of as infinite series which do converge pointwise. In practice, and in this text, only finite sums are actually computed. Now this simple splitting is likely to yield poor computational results because it can introduce artificial discontinuities. For example, consider the trivial problem

$$\Delta u = 0 \qquad \text{in } D$$
$$u = 1 \qquad \text{on } \partial D$$

which has the unique solution $u(x, y) \equiv 1$.

The corresponding problems for u_1 and u_2 are

$$\Delta u_1 = 0$$
$$u_1 = 0 \quad \text{on } x = 0 \text{ and } x = a$$
$$u_1 = 1 \quad \text{on } y = 0 \text{ and } y = b,$$
$$\Delta u_2 = 0$$
$$u_2 = 1 \quad \text{on } x = 0 \text{ and } x = a$$
$$u_2 = 0 \quad \text{on } y = 0 \text{ and } y = b.$$

Hence u_1 is the solution of a Dirichlet problem with discontinuous boundary data, and while we can formally compute an approximate solution, we will have to contend with a Gibbs phenomenon near the corner points. An identical problem arises in the computation of u_2.

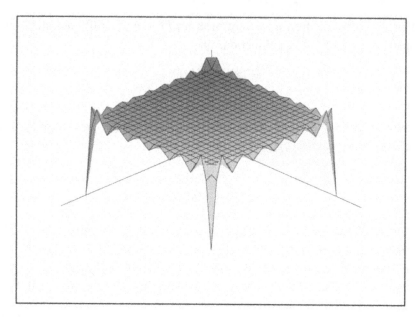

Figure 8.1: (a) Solution u_{30} of $\Delta u = 0$, $u = 1$ on ∂D, obtained with a formal splitting.

Fig. 8.1a shows a plot of the solution $u_N = u_{1N} + u_{2N}$ obtained with the formal splitting (8.2) and (8.3). This is, of course, a terrible approximation of the analytic solution $u = 1$. In general, the formal splitting discussed above will succeed and provide a reasonable solution but may require many terms in the approximating sum in order to squeeze the effect of the Gibbs phenomenon into a small region near the corners of D. The approximation becomes even more suspect should derivatives of u_N be needed.

When g is continuous on ∂D in (8.1), then we can avoid artificial discontinuities at the corners of D by preconditioning the original problem. We write

$$w(x, y) \equiv u(x, y) - v(x, y)$$

where v is any smooth function which takes on the values of g at the corners of D. A simple choice is the polynomial

$$v(x, y) = c_0 + c_1 x + c_2 y + c_3 xy$$

which satisfies Laplace's equation. A little algebra yields

$$v(x, y) = g(0, 0)\frac{(a - x)(b - y)}{ab} + g(a, 0)\frac{x(b - y)}{ab}$$
$$+ g(a, b)\frac{xy}{ab} + g(0, b)\frac{y(a - x)}{ab} \, . \tag{8.4}$$

When problem (8.1) is rewritten for w, then a Dirichlet problem like (8.1) results where the boundary data are zero at the corners of ∂D. Now we can apply a formal splitting. We write

$$w = w_1 + w_2$$

where

$$\Delta w_1 = F_1(x, y)$$

$$w_1(0, y) = w_1(a, y) = 0$$

$$w_1(x, 0) = g(x, 0) - v(x, 0), \quad w_1(x, b) = g(x, b) - v(x, b),$$

$$\Delta w_2 = F_2(x, y)$$

$$w_2(0, y) = g(0, y) - v(0, y), \quad w_2(a, y) = g(a, y) - v(a, y)$$

$$w_2(x, 0) = w_2(x, b) = 0.$$

Two Dirichlet problems with continuous boundary data and hence continuous solutions result. Since $w_1(0, 0) = w_1(a, 0) = w_1(0, b) = w_1(a, b) = 0$, no Gibbs phenomenon will arise when $w_1(x, 0)$ and $w_1(x, b)$ are projected into span$\{\phi_m(x)\}$. Similarly, the projections of $w_2(0, y)$ and $w_2(a, y)$ into span$\{\psi_n(y)\}$ will converge uniformly.

We view the transformation of (8.1) into a new Dirichlet problem with vanishing boundary data at the corners of ∂D as a preconditioning of the original problem.

For illustration we show in Figs. 8.1b,c the approximate solutions obtained from a formal splitting without preconditioning and from a splitting after preconditioning with (8.4) for the problem

$$\Delta u = 0, \qquad (x, y) \in D = (0, 1) \times (0, 1)$$

$$u = xy|(1 - 2x)(1 - 3y)|, \qquad (x, y) \in \partial D.$$

Note that the boundary function g is continuous on D but not differentiable. We do not know an analytic solution for this problem, but we do know from Section 1.3 that it has a unique classical solution.

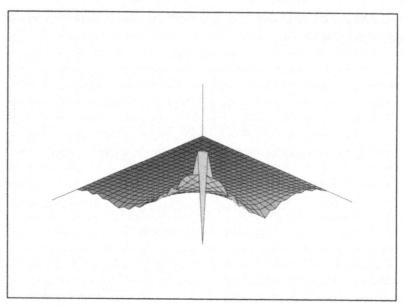

Figure 8.1: (b) Solution u_{20} of $\Delta u = 0$ in D, $u = xy|(1 - 2x)(1 - 3y)|$ on ∂D obtained with a formal splitting.

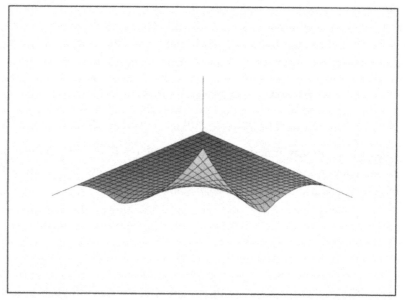

Figure 8.1: (c) Solution u_{20} obtained after preconditioning with (8.4).

Example 8.2 Preconditioning for general boundary data.

Separation of variables is applicable to Poisson's equation

$$\Delta u = F(x,y), \qquad (x,y) \in D = (0,a) \times (0,b) \tag{8.5a}$$

subject to boundary data of the third kind

$$c_1(x,y)\frac{\partial u}{\partial n} + c_2(x,y)u = g(x,y), \qquad (x,y) \in \partial D \tag{8.5b}$$

where $\frac{\partial u}{\partial n}$ denotes the outward normal derivative of u on ∂D (e.g., $\frac{\partial u}{\partial n}(0,y) = -\frac{\partial u}{\partial x}(0,y)$) and where c_1 and c_1 are nonnegative *constants* on the line segments $x = 0, a$ and $y = 0, b$ of ∂D. At the corner points the boundary condition is not defined and not needed for the approximate solution. We do assume that g is continuous on each line segment and has limits as we approach the corners.

If $g(0,y)$ and $g(a,y)$ are twice continuously differentiable, then it is straightforward to find a smooth function v defined on D which assumes the given boundary conditions at $x = 0$ and $x = a$. We write

$$v(x,y) = g(0,y)f_1(x) + g(a,y)f_2(x)$$

where f_1 and f_2 are smooth functions which satisfy

$$-c_1(0,y)f_1'(0) + c_2(0,y)f_1(0) = 1$$
$$c_1(a,y)f_1'(a) + c_2(a,y)f_1(a) = 0$$

and

$$-c_1(0,y)f_2'(0) + c_2(0,y)f_2(0) = 0$$
$$c_1(a,y)f_2'(a) + c_2(a,y)f_2(a) = 1.$$

A simple choice would be functions of the form

$$f_1(x) = k_{11}(x - a) + k_{12}(x - a)^2$$
$$f_2(x) = k_{21}x + k_{22}x^2$$

for appropriate constants k_{ij} depending on c_1 and c_2. When we define

$$w(x,y) = u(x,y) - v(x,y),$$

then problem (8.5) is transformed to

$$\Delta w = F(x,y) - \Delta v(x,y) \qquad \text{in } D$$

$$c_1\frac{\partial w}{\partial n} + c_2 w = \begin{cases} 0, & x = 0, a \\ g(x,y) - \left[c_1\frac{\partial v}{\partial n} + c_2 v\right], & y = 0, b. \end{cases}$$

With homogeneous boundary conditions on opposing sides of ∂D we can apply the eigenfunction expansion of Chapter 5. Note that the computed solution $u_N = w_N + v$ will satisfy the given boundary conditions on $x = 0$ and $x = a$ exactly and the approximate boundary condition

$$c_1 \frac{\partial u_N}{\partial n} + c_2 u_N = P_N g(x,y) - P_N \left(c_1 \frac{\partial v}{\partial n} + c_2 v \right) + \left(c_1 \frac{\partial v}{\partial n} + c_2 v \right) \quad \text{on } y = 0, b.$$

This approach breaks down when the function $v(x,y)$ defined above (and its analogue which interpolates the data $g(x,0)$ and $g(x,b)$) is not differentiable. Then, in principle, problem (8.5) can be solved with the formal splitting

$$u = u_1 + u_2$$

where

$$\Delta u_1 = F_1(x,y) \qquad (x,y) \in D$$

$$c_1(x,y) \frac{\partial u_1}{\partial n} + c_2(x,y) u_1 = \begin{cases} 0 & \text{for } y \in (0,b) \text{ and } x = 0, a \\ g(x,y) & \text{for } x \in (0,a) \text{ and } y = 0, b, \end{cases}$$

$$\Delta u_2 = F_2(x,y) \qquad (x,y) \in D$$

$$c_1(x,y) \frac{\partial u_2}{\partial n} + c_2(x,y) u_2 = \begin{cases} g(x,y) & \text{for } y \in (0,b) \text{ and } x = 0, a \\ 0 & \text{for } x \in (0,a) \text{ and } y = 0, b, \end{cases}$$

and

$$F(x,y) = F_1(x,y) + F_2(x,y).$$

As in Example 8.1, this formal splitting can introduce discontinuities into the boundary data at the corners of ∂D which may not be present in the original problem (8.5). In this case a preconditioning of the problem is advantageous which assures that the splitting has continuous boundary data. The exposition of preconditioning for the general case is quite involved and included here for reference. It presupposes that g is such that the problem (8.5) admits a classical solution with high regularity. This requires certain consistency conditions on g at the corners of ∂D which are derived below.

To be specific let us examine the solution u of (8.5) near the corner point (a,b). We wish to find a smooth function $v(x,y)$ so that the new dependent variable

$$w \equiv u - v,$$

which solves the problem

$$\Delta w = F - \Delta v \equiv G(x,y)$$

$$c_1(x,y)\frac{\partial w}{\partial n} + c_2(x,y)w = h(x,y),$$

can be found with a formal splitting without introducing an artificial singularity at (a, b).

Suppose for the moment that such a v has been found. If we write

$$w = w_1 + w_2 \quad \text{and} \quad G = G_1 + G_2,$$

then w_1 and w_2 solve

$$\Delta w_1 = G_1(x,y) \qquad (x,y) \in D$$

$$c_1(x,y)\frac{\partial w_1}{\partial n} + c_2(x,y)w_1 = \begin{cases} 0 & \text{for } y \in (0,b) \text{ and } x = 0,a \\ h(x,y) & \text{for } x \in (0,a) \text{ and } y = 0,b, \end{cases}$$

and

$$\Delta w_2 = G_2(x,y) \qquad (x,y) \in D$$

$$c_1(x,y)\frac{\partial w_2}{\partial n} + c_2(x,y)w_2 = \begin{cases} h(x,y) & \text{for } y \in (0,b) \text{ and } x = 0,a \\ 0 & \text{for } x \in (0,a) \text{ and } y = 0,b \end{cases}$$

where

$$h(x,y) = g(x,y) - \left[c_1(x,y)\frac{\partial v}{\partial n}(x,y) + c_2(x,y)v(x,y) \right]. \qquad (8.6)$$

Now we can apply our eigenfunction expansion. The eigenvalue problem associated with w_1 is

$$\phi''(x) = \mu\phi(x)$$

$$-c_1(0,y)\phi'(0) + c_2(0,y)\phi(0) = 0$$

$$c_1(a,y)\phi'(a) + c_2(a,y)\phi(a) = 0.$$

Its solution is discussed at length in Section 3.1. Let $\{-\lambda_m^2\}$ and $\{\phi_m(x)\}$ denote the eigenvalues and eigenfunctions for $m \geq m_0$ where, depending on c_1 and c_2, $m_0 = 0$ or 1. Then the problem for w_1 is approximated by

$$\Delta w_1 = P_M G_1(x,y)$$

$$c_1(x,y)\frac{\partial w_1}{\partial n} + c_2(x,y)w_1 = \begin{cases} 0 & \text{for } x = 0,a \\ P_M h(x,y) & \text{for } y = 0,b, \end{cases}$$

where

$$P_M G_1(x,y) = \sum_{m=m_0}^{M} \gamma_m(y)\phi_m(x)$$

$$\gamma_m(y) = \frac{\langle G_1(x,y), \phi_m(x)\rangle}{\langle \phi_m, \phi_m \rangle}, \qquad \langle f,g \rangle = \int_0^a f(x)g(x)dx$$

$$P_M h(x,0) = \sum_{m=m_0}^{M} \hat{\alpha}_m \phi_m(x)$$

$$\hat{\alpha}_m = \frac{\langle h(x,0), \phi_m(x) \rangle}{\langle \phi_m, \phi_m \rangle},$$

and

$$P_M h(x,b) = \sum_{m=m_0}^{M} \hat{\beta}_m \phi_m(x).$$

$$\hat{\beta}_m = \frac{\langle h(x,b), \phi_m(x) \rangle}{\langle \phi_m, \phi_m \rangle}.$$

This problem is solved by

$$w_{1M}(x,y) = \sum_{m=m_0}^{M} \alpha_m(y)\phi_m(x)$$

where

$$-\lambda_m^2 \alpha_m(y) + \alpha_m''(y) = \gamma_m(y)$$

$$-c_1(x,0)\alpha_m'(0) + c_2(x,0)\alpha_m(0) = \hat{\alpha}_m$$

$$c_1(x,b)\alpha_m'(b) + c_2(x,b)\alpha_m(b) = \hat{\beta}_m.$$

The solution $\alpha_n(y)$ has the form

$$\alpha_n(y) = d_1 \sinh \lambda_n y + d_2 \cosh \lambda_n y + \alpha_{np}(y).$$

Whether one can solve for the coefficients d_1 and d_2 depends on the data of the problem. For $c_1 c_2 > 0$ any solution we can construct is necessarily unique because the solution of the Robin problem is unique as shown in Section 1.3.

Similarly, w_2 is approximated by the solution w_{2N} of the problem

$$\Delta w_2 = P_N G_2(x,y)$$

$$c_1(x,y)\frac{\partial w_2}{\partial n} + c_2(x,y)w_2 = \begin{cases} P_N h(x,y) & \text{for } x = 0, a \\ 0 & \text{for } y = 0, b, \end{cases}$$

where

$$P_N G_2(x,y) = \sum_{n=n_0}^{N} \gamma_n(x)\psi_n(y),$$

$$\gamma_n(x) = \frac{\langle G_2(x,y), \psi_n \rangle}{\langle \psi_n, \psi_n \rangle} \qquad \langle f, g \rangle = \int_0^b f(y)g(y)dy$$

$$P_N h(0,y) = \sum_{n=n_0}^{N} \hat{\alpha}_n \psi_n(y)$$

$$\hat{\alpha}_n = \frac{\langle h(0,y), \psi_n(y) \rangle}{\langle \psi_n, \psi_n \rangle}$$

and

$$P_N h(a,y) = \sum_{n=n_0}^{N} \hat{\beta}_n \psi_n(y)$$

$$\hat{\beta}_n = \frac{\langle h(a,y), \psi_n(y) \rangle}{\langle \psi_n, \psi_n \rangle}.$$

Here $\{-\delta_n^2\}$ and $\{\psi_n(y)\}$ are the eigenvalues and eigenfunctions of

$$\psi''(y) = -\delta^2 \psi(y)$$

$$-c_1(x,0)\psi'(0) + c_2(x,0)\psi(0) = 0$$

$$c_1(x,b)\psi'(b) + c_2(x,b)\psi(b) = 0.$$

The solution is

$$w_{2N}(x,y) = \sum_{n=n_0}^{N} \alpha_n(x)\psi_n(y)$$

where

$$-\delta_n^2 \alpha_n(x) + \alpha_n''(x) = \gamma_n(x)$$

$$-c_1(0,y)\alpha_n'(0) + c_2(0,y)\alpha_n(0) = \hat{\alpha}_n$$

$$c_1(a,y)\alpha_n'(a) + c_2(a,y)\alpha_n(a) = \hat{\beta}_n.$$

Let us now turn to the choice of v. The discussion of the Gibbs phenomenon in Chapter 4 suggests that the preconditioning should result in boundary data that satisfy the same homogeneous boundary condition at the corner points as the eigenfunctions used for their approximation. This will be the case for $h(x,b)$ and $h(a,y)$ at the point (a,b) if

$$\lim_{x \to a} [c_1(a,y)h_x(x,b) + c_2(a,y)h(x,b)] = 0$$

and

$$\lim_{y \to b} [c_1(x,b)h_y(a,y) + c_2(x,b)h(a,y)] = 0.$$

(Remember that the coefficients $c_i(x,y)$ are constants along $x = 0, a$ and $y = 0, b$.)

Substitution of (8.6) for $h(x,y)$ leads to the two equations

$$\lim_{x \to a} [c_1(a,y)[g_x(x,b) - \{c_1(x,b)v_{yx}(a,b) + c_2(x,b)v_x(a,b)\}]$$

$$+ c_2(a,y)[g(x,b) - \{c_1(x,b)v_y(a,b) + c_2(x,b)v(a,b)\}] = 0$$

and

$$\lim_{y \to a}[c_1(x,b)[g_y(a,y) - \{c_1(a,y)v_{xy}(a,b) + c_2(a,y)v_y(a,b)\}]$$
$$+ c_2(x,b)[g(a,y) - \{c_1(a,y)v_x(a,b) + c_2(a,y)v(a,b)\}] = 0.$$

Thus

$$c_1(a,y)c_1(x,b)v_{xy}(a,b) + c_1(a,y)c_2(x,b)v_x(a,b) \tag{8.7}$$
$$+ c_1(x,b)c_2(a,y)v_y(a,b) + c_2(a,y)c_2(x,b)v(a,b)$$
$$= \lim_{x \to a}[c_1(a,y)g_x(x,b) + c_2(a,y)g(x,b)],$$
$$c_1(a,y)c_1(x,b)v_{xy}(a,b) + c_1(a,y)c_2(x,b)v_x(a,b)$$
$$+ c_1(x,b)c_2(a,y)v_y(a,b) + c_2(a,y)c_2(x,b)v(a,b)$$
$$= \lim_{y \to b}[c_1(x,b)g_y(a,y) + c_2(x,b)g(a,y)].$$

Both equations have the same left-hand side and hence can have a solution only if

$$\lim_{x \to a}[c_1(a,y)g_x(x,b) + c_2(a,y)g(x,b)] \tag{8.8}$$
$$= \lim_{y \to b}[c_1(x,b)g_y(a,y) + c_2(x,b)g(a,y)].$$

We shall call the boundary condition consistent at (a,b) if equation (8.8) holds. The meaning of (8.8) becomes a little clearer if we look at two special cases. Suppose that Dirichlet data are imposed on $x = a$ and $y = b$, i.e.

$$u(x,y) = g(x,y) \quad \text{on } x = a \text{ and on } y = b.$$

Then $c_1(a,y) = c_1(x,b) = 0$ and $c_2(a,y) = c_2(x,b) = 1$. The consistency condition (8.8)

$$\lim_{y \to b} g(a,y) = \lim_{x \to a} g(x,b)$$

simply implies that g is continuous on ∂D at (a,b). Equation (8.7) reduces to the preconditioning condition known from Example 8.1

$$v(a,b) \doteq g(a,b).$$

Suppose next that

$$\frac{\partial u}{\partial n} = g(x,y) \quad \text{on } x = a \text{ and on } y = b.$$

Then $c_1 = 1$ and $c_2 = 0$ and (8.8) requires that

$$\lim_{x \to a} g_x(x,b) = \lim_{y \to b} g_y(a,y).$$

Hence if g is differentiable on ∂D at (a, b), then $u_{xy}(x, y)$ is continuous on ∂D at (a, b). In this case (8.7) requires

$$v_{xy}(a, b) = \lim_{x \to a} g_x(x, b).$$

In general it is straightforward to show with Gaussian elimination that if (8.5b) holds and can be differentiated along $x = a$ and $y = b$, then the resulting four equations for $u(a, b)$, $u_x(a, b)$, $u_y(a, b)$, and $u_{xy}(a, b)$ are overdetermined and consistent only if (8.8) holds. If the data are not consistent, then preconditioning is generally not possible. On the other hand, if this consistency condition is met, then we can choose for $v(x, y)$ any smooth function which satisfies (8.7). Similar arguments apply to $h(x, y)$ at the other three corners of the rectangle. If the data are consistent everywhere, then we look for a smooth function $v(x, y)$ which satisfies four equations like (8.7). It is straightforward to verify that the function

$$v_{ab}(x, y) = \frac{A}{B} x^2 y^2$$

with

$$A = \lim_{x \to a} [c_1(a, y) g_x(x, b) + c_2(a, y) g(x, b)]$$

and

$$B = 4 c_1(a, y) c_1(x, b) ab + 2 c_1(a, y) c_2(x, b) ab^2 + 2 c_1(x, b) c_2(a, y) a^2 b$$
$$+ c_2(a, y) c_2(x, b) a^2 b^2$$

satisfies equation (8.7) and

$$v_{ab} = v_{abx} = v_{aby} = v_{abxy} = 0 \quad \text{at } (0, 0), \ (a, 0), \ \text{and} \ (0, b).$$

Similar functions of the form

$$v_{00}(x, y) = C(x - a)^2 (y - b)^2$$

$$v_{0b}(x, y) = D(x - a)^2 y^2$$

$$v_{a0}(x, y) = E x^2 (y - b)^2$$

with appropriate coefficients D, D, and E allow us to express $v(x, y)$ as

$$v(x, y) = v_{00}(x, y) + v_{a0}(x, y) + v_{ab}(x, y) + v_{0b}(x, y).$$

Preconditioning with such a polynomial introduces an additional smooth source term into Poisson's equation for w, i.e.

$$\Delta w = \Delta u - \Delta v = F(x, y) - \Delta v(x, y),$$

but its influence on the eigenfunction expansion solution appears to be small because the problems to be solved with the formal splitting have classical solutions with the same smoothness as the solution u of the original problem. Hence preconditioning of the data to ensure that the subsequent splitting has a smooth solution is not particularly arduous and greatly improves the accuracy of the approximate solution.

As illustration let us consider the problem

$$\Delta u = 0, \qquad (x, y) \in D = (0, 1) \times (0, 1)$$

$$u_y(x, 0) = 0, \qquad x \in (0, 1)$$

$$u(1, y) = \frac{y^2}{2}, \qquad y \in (0, 1)$$

$$u_y(x, 1) = x(2x - 1), \qquad x \in (0, 1)$$

$$u(0, y) = (1 - y)^2, \qquad y \in (0, 1).$$

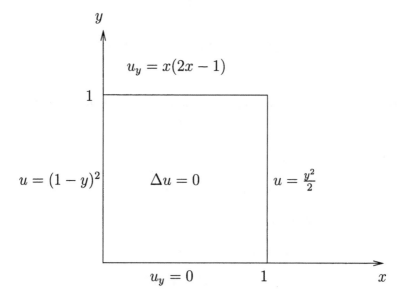

Figure 8.2: (a) A graphical display of the test problem.

A useful mnemonic to visualize the problem and help keep track of various splittings is to graph geometry and problem as in Fig. 8.2a. It is straightforward to check that these data are consistent at the corners $(1, 0)$, $(1, 1)$, and $(0, 1)$ but not at $(0, 0)$ where

$$\lim_{y \to 0} u_y(0, y) \neq \lim_{x \to 0} u_y(x, 0).$$

We shall solve this problem three different ways. First, we write

$$w(x, y) = u(x, y) - v(x, y)$$

with

$$v(x, y) = (1 - y)^2(1 - x) + xy^2/2, \tag{8.9}$$

and apply the eigenfunction expansion method to

$$\Delta w = x - 2 \equiv G(x, y)$$

$$w(0, y) = w(1, y) = 0$$

$$w_y(x, 0) = 2(1 - x), \qquad w_y(x, 1) = 2x(x - 1).$$

The eigenfunctions are $\phi_n(x) = \sin \lambda_n x$, $\lambda_n = n\pi$, $n \geq 1$ so that we will see a Gibbs phenomenon in the approximation of $G(x, y)$ at $x = 0$ and $x = 1$, and in the approximation of $w_y(x, 0)$ at $x = 0$. Our second solution is obtained from the formal splitting $u = u_1 + u_2$ where

$$\Delta u_1 = 0$$

$$u_1(0, y) = u_1(1, y) = 0$$

$$u_{1y}(x, 0) = 0, \qquad u_{1y}(x, 1) = x(2x - 1),$$

and

$$\Delta u_2 = 0$$

$$u_2(0, y) = (1 - y)^2, \qquad u_2(1, y) = \frac{y^2}{2}$$

$$u_{2y}(x, 0) = u_{2y}(x, 1) = 0.$$

It is easy to see that the boundary data for u_1 are not consistent at $(1, 1)$ and that those for u_2 are not consistent at $(0, 0)$ and $(1, 1)$. The discontinuity in u_{2y} at $(0, 0)$ is inherent in the problem. The discontinuities at $(1, 1)$ are caused by the formal splitting.

The artificial inconsistency at $(1, 1)$ can be removed by preconditioning the data. We observe that the eigenfunctions $\phi(x)$ and $\psi(y)$ of the formal splitting satisfy the boundary conditions

$$\phi(0) = \phi(1) = 0$$

$$\psi'(0) = \psi'(1) = 0.$$

We define

$$w = u - v.$$

Then v should be chosen such that the boundary condition $w_y(x,1) = x(2x-1) - v_y(x,1)$ can be expanded in terms of $\{\phi_m(x)\}$ without introducing a Gibbs phenomenon. This requires that $w_y(0,1) = 0$ and $w_y(1,1) = 0$. Similarly

$$w(0,y) = (1-y)^2 - v(0,y)$$

should satisfy the boundary conditions of $\{\psi_n(y)\}$, i.e.

$$w_y(0,0) = w_y(0,1) = 0.$$

Similar expressions hold along the lines $y = 0$ and $x = 1$. Altogether we have the conditions

$$\lim_{x \to 0} w_y(x,1) = \lim_{x \to 1} w_y(x,1) = 0$$

$$\lim_{x \to 0} w_y(x,0) = \lim_{x \to 1} w_y(x,0) = 0$$

$$\lim_{y \to 0} w_y(0,y) = \lim_{y \to 1} w_y(0,y) = 0$$

$$\lim_{y \to 0} w_y(1,y) = \lim_{y \to 1} w_y(1,y) = 0.$$

This leads to

$$v_y(1,0) = v_y(0,1) = 0, \qquad v_y(1,1) = 1$$

and the inconsistent condition

$$\lim_{y \to 0} v_y(0,y) = -2, \qquad \lim_{x \to 0} v_y(x,0) = 0.$$

We observe that the function

$$v(x,y) = \frac{xy^2}{2}$$

satisfies the conditions at $(1,0)$, $(1,1)$, and $(0,1)$. Then the preconditioned problem is

$$\Delta w = -x$$

$$w(0,y) = (1-y)^2, \qquad w(1,y) = 0$$

$$w_y(x,0) = 0, \qquad w_y(x,1) = 2x(x-1).$$

We split the problem by writing

$$w = w_1 + w_2$$

where

$$\Delta w_1 = 0$$

$$w_1(0,y) = w_1(1,y) = 0$$

$$w_{1y}(x, 0) = 0, \qquad w_{1y}(x, 1) = 2x(x - 1)$$

and

$$\Delta w_2 = -x$$
$$w_2(0, y) = (1 - y)^2, \qquad w_2(1, y) = 0$$
$$w_{2y}(x, 0) = w_{2y}(x, 1) = 0.$$

We verify by differentiating the Dirichlet data along the boundary that the data for w_1 and w_2 are consistent at the corners except for the point $(0, 0)$ where

$$\lim_{y \to 0} w_{2y}(0, y) \neq \lim_{x \to 0} w_{2y}(x, 0).$$

The separation of variables approximations to w_1 and w_2 are readily obtained. Since the preconditioning in this problem was carried out to preserve the smoothness of u_y at three corners, we show in Figs. 8.2b, c, d the surfaces for u_y obtained with the three formulations. The discontinuity at $(0, 0)$ is unavoidable, but otherwise one would expect a good solution to be smooth. Clearly it pays to precondition the problem before splitting it. Zeroing out inhomogeneous but smooth boundary data on opposite sides with (8.9) and using an eigenfunction expansion preserves smoothness, requires no preconditioning, and is our choice for such problems.

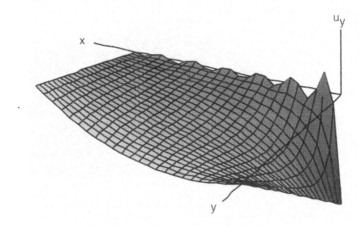

Figure 8.2 (b) Eigenfunction expansion with exact boundary data at $x = 0, 1$ and a Gibbs phenomenon on $y = 0$.

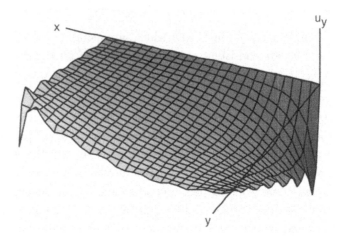

Figure 8.2 (c) Formal splitting with singularity at $(1, 1)$.

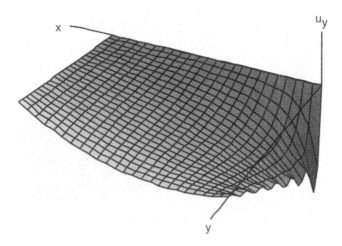

Figure 8.2 (d) Solution u_y from a formal splitting after preconditioning, with a Gibbs phenomenon on $x = 0$.

Example 8.3 Poisson's equation with Neumann boundary data.

The problem

$$\Delta u = F(x, y), \qquad (x, y) \in D = (0, a) \times (0, b)$$

$$\frac{\partial u}{\partial n} = g(x, y), \qquad (x, y) \in \partial D$$

is examined in Section 1.3. We know that a solution can exist only if

$$\int_D F \, dx \, dy = \oint_{\partial D} g \, ds, \qquad (8.10)$$

and we shall assume henceforth that this condition holds.

For the eigenfunction approach we need homogeneous boundary data on opposite sides of ∂D. If $g(0, y)$ and $g(a, y)$ are smooth, then for

$$v(x, y) = g(0, y) \frac{(a - x)^2}{2a} + g(a, y) \frac{x^2}{2a}$$

the new dependent variable

$$w(x, y) = u(x, y) - v(x, y)$$

solves

$$\Delta w = F(x, y) - \Delta v(x, y)$$

$$w_x(0, y) = w_x(a, y) = 0$$

$$-w_y(x, 0) = g(x, 0) + v_y(x, 0)$$

$$w_y(x, b) = g(x, b) - v_y(x, b).$$

Since for any smooth function h the divergence theorem

$$\int_D h \, dx \, dy = \oint_{\partial D} \frac{\partial h}{\partial n} \, ds$$

holds, we see that the new problem satisfies the consistency condition (8.10). Furthermore, since the eigenfunctions associated with this Neumann problem are $\cos \lambda_n x$, $\lambda_n = \frac{n\pi}{a}$, $n = 0, 1, \ldots$, we find by integrating that

$$\int_D P_N F(x, y) dx \, dy = \sum_{n=0}^{N} \int_0^b \gamma_n(y) dy \int_0^a \phi_n(x) dx$$

$$= a \int_0^b \gamma_0(y) dy = \int_D F(x, y) dx \, dy$$

and

$$\int_0^a P_N g(x,0)dx = \hat{\alpha}_0 a = \int_0^a g(x,0)dx$$

so that the approximating problem obtained by projecting F and g also satisfies the consistency condition (8.10). Hence the Neumann problem is solvable in terms of an eigenfunction expansion.

If $\{g(0,y), g(a,y)\}$ and $\{g(x,0), g(x,b)\}$ are not twice differentiable, then Δv does not exist and we are forced into a formal splitting of the Neumann problem. Such a splitting requires that each new problem satisfy its own compatibility condition analogous to (8.10). We can force compatibility by writing, for example

$$\Delta u_1 = F_1(x,y)$$

$$u_{1x}(0,y) = u_{1x}(a,y) = 0$$

$$-u_{1y}(x,0) = g(x,0), \qquad u_{1y}(x,b) = g(x,b),$$

$$\Delta u_2 = F_2(x,y)$$

$$-u_{2y}(x,0) = u_{2y}(x,b) = 0$$

$$-u_{2x}(0,y) = g(0,y), \qquad u_{2x}(a,y) = g(a,y)$$

where F_1 is chosen such that

$$\int_D F_1(x,y)dx\,dy = \int_0^b [g(x,0) + g(x,b)]dx$$

and

$$F_1(x,y) + F_2(x,y) = F(x,y).$$

A possible choice for F_1 is

$$F_1(x,y) = F(x,y) - A$$

where

$$A = \frac{1}{ab}\int_0^b [g(0,y) + g(a,y)]dy.$$

An eigenfunction solution for u_1 and u_2 is straightforward to compute.

When the boundary data of the original problem are consistent in the sense of equation (8.8), i.e., at (a,b)

$$\lim_{x\to a} g_x(x,b) = \lim_{y\to b} g_y(a,y),$$

then the above splitting should be applied to $w = u - v$ after preconditioning with an appropriate v. Since v is a given smooth function, it follows automatically that $\left(\Delta v, \frac{\partial v}{\partial n}\right)$ are compatible in the sense of (8.10). Hence preconditioning and the source-flux balance (8.10) are logically independent.

Example 8.4 A discontinuous potential.

We consider a long rectangular metal duct whose bottom and sides are grounded, and whose top is insulated against the sides and held at a constant nonzero voltage. We wish to find the electrostatic potential inside the duct. We shall assume that far enough away from the ends of the duct the potential is essentially two dimensional and described by the following mathematical model:

$$\Delta u = 0 \qquad \text{in } D = (0,2) \times (0,1)$$

$$u(0,y) = u(x,0) = u(2,y) = 0, \qquad (x,y) \in \partial D$$

$$u(x,1) = 40, \qquad x \in (0,2).$$

This problem is straightforward to solve with separation of variables and is chosen only to illustrate that preconditioning is not always possible and to demonstrate the influence of the Gibbs phenomenon on the computed answers.

The associated eigenvalue problem is

$$\phi''(x) = \mu\phi(x)$$

$$\phi(0) = \phi(2) = 0$$

with solutions

$$\phi_n(x) = \sin\lambda_n x, \qquad \lambda_n = \frac{n\pi}{2}, \qquad n = 1,\dots.$$

The approximate solution

$$u_N(x,y) = \sum_{n=1}^{N} \alpha_n(y)\phi_n(x)$$

solves

$$-\lambda_n^2 \alpha_n(y) + \alpha_n''(y) = 0$$

$$\alpha_n(0) = 0, \qquad \alpha_n(1) = \frac{40\langle 1, \phi_n\rangle}{\langle\phi_n, \phi_n\rangle}.$$

The solution is

$$u_N(x,y) = \frac{80}{\pi}\sum_{n=1}^{N} \frac{[1-(-1)^n]}{n}\frac{\sinh\lambda_n y}{\sinh\lambda_n}\sin\lambda_n x. \qquad (8.11)$$

The discussion of the Gibbs phenomenon in Chapter 4 applies to the Fourier sine series of $u(x,1) = 40$. We can infer that for all $x \in (0,2)$

$$u_N(x,1) \to 40 \qquad \text{as } N \to \infty$$

but also that for sufficiently large N we will have the approximation error

$$\sup_{x \in (0,2)} u_N(x,1) - 40 \cong .0894 * 80 \cong 7.15.$$

It also follows from (8.11) that for all n

$$\left| \frac{80}{\pi} \frac{[1-(-1)^n]}{n} \frac{\sinh \lambda_n y}{\sinh \lambda_n} \sin \lambda_n x \right| \le \frac{160}{n\pi} \frac{1}{1-e^{-\pi}} \left(\frac{e^{\frac{\pi y}{2}}}{e^{\frac{\pi}{2}}} \right)^n \equiv K \frac{\alpha^n(y)}{n}$$

where $K \cong 53.23$ and

$$\alpha(y) = \left(\frac{e^{\frac{\pi y}{2}}}{e^{\frac{\pi}{2}}} \right) < 1 \quad \text{for } y < 1.$$

Hence for all $y \le y_0 < 1$ the approximation u_N converges uniformly on $[0,2] \times [0, y_0]$ to a function $u(x,y)$ as $N \to \infty$. If we accept $u(x,y)$ as the (generalized) solution of our problem, then

$$|u_N(x,y) - u(x,y)| \le K \sum_{n=N+1}^{\infty} \frac{\alpha^n}{n} \le K \int_N^{\infty} \frac{e^{x \ln \alpha}}{x} dx = KE_1(-N \ln \alpha),$$

where $E_1(z)$ is the exponential integral

$$E_1(z) = \int_1^{\infty} \frac{e^{-zt}}{t} dt, \quad z > 0.$$

Hence to guarantee an error bound of 10^{-6} at a point (x,y), $y < 1$, it is sufficient to find N such that

$$53.23 E_1(-N \ln \alpha) \cong 10^{-6} \quad \text{or} \quad E_1(-N \ln \alpha) \cong 1.88 \; 10^{-8}.$$

The computer tells us that

$$E_1(15.015) \sim 1.88 \; 10^{-8}$$

so that

$$N \cong \frac{15.015}{-\ln \alpha} = \frac{30.3}{\pi(1-y)}. \tag{8.12}$$

Of course, this estimate is pessimistic but, as Table 8.4 shows, still predicts the correct order for the number of terms to give an error of $\le 10^{-6}$.

Table 8.4 Numerical solution of the potential problem

Coordinates	Solution	N Predicted by (8.12)	N Required
(1.999, .999)	20.0002	9559	5615
(1, .5)	17.8046	20	19
(.001, .001)	$6.19008 \; 10^{-5}$	10	3

The analytic solution u is taken to be u_N for $N = 10,000$ which is guaranteed to differ from the infinite series solution by less than 10^{-6} (in infinite precision arithmetic). The required N is the smallest N for which we observe $|u - u_N| \leq 10^{-6}$ at the given point.

As we have seen before, we may need hundreds and thousands of terms in our approximate solution near points where the analytic solution has steep gradients.

Finally, we observe that the (approximate) equipotential lines in the duct are the level curves of

$$u_N(x, y) = c.$$

The electric field lines are perpendicular to these contours. Since $u_N(x, y)$ is an analytic solution of Laplace's equations, one knows from complex variable theory that the field lines are the level curves of the harmonic conjugate v_N of u_N which is found by solving the differential equations

$$v_{Ny} = u_{Nx}, \qquad v_{Nx} = -u_{Ny}.$$

From (8.11) now follows that up to an additive constant

$$v_N(x, y) = \frac{80}{\pi} \sum_{n=1}^{N} \frac{[1 - (-1)^n]\cosh \lambda_n y}{\sinh \lambda_n} \cos \lambda_n x.$$

Equipotential and field lines in the duct are shown in Fig. 8.4.

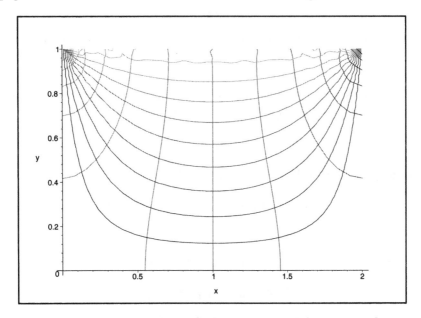

Figure 8.4: Equipotential and field lines obtained from u_{20} and v_{20}.

Example 8.5 Lubrication of a plane slider bearing.

Reynolds equation for the incompressible lubrication of bearings can some-times be solved with separation of variables. One such application is the calcula-tion of the lubricant film pressure in a plane slider bearing. It leads to the follow-ing mathematical model: suppose the bearing has dimensions $D = (0, a) \times (0, b)$. Then the pressure in the lubricant is given by Reynolds equation

$$(h^3 p_x)_x + (h^3 p_z)_z = c h_x, \qquad (x, z) \in D$$

$$p = 0, \qquad (x, z) \in D$$

where h is the film thickness and c is a known constant. (For a description of the model see, e.g., [17].)

A closed form solution of this problem is possible only for special film shapes. If the film thickness depends on x only, then separation of variables applies. Assuming that $h(x) > 0$ we can rewrite Reynolds equation in the form

$$p_{xx} + 3 \frac{h'(x)}{h(x)} p_x + p_{zz} = \frac{ch'(x)}{h(x)^3} .$$

Following Section 3.2 we associate with it the regular Sturm–Liouville eigenvalue problem

$$(h^3(x)\phi')' = \mu h^3(x)\phi$$

$$\phi(0) = \phi(a) = 0.$$

Standard arguments show that any eigenvalue must be real and negative, and that eigenfunctions corrresponding to distinct eigenvalues are orthogonal in $L_2(0, a, h^3)$.

Eigenvalues and eigenfunctions can be found, at least numerically, when the film thickness is given by the linear profile

$$h(x) = A + Bx, \qquad A, B > 0.$$

Then the eigenvalue problem can be written in the form

$$(A + Bx)^2 \phi'' + 3B(A + Bx)\phi' = \mu(A + Bx)^2 \phi$$

$$\phi(0) = \phi(a) = 0.$$

With the change of variable $y = A + Bx$ it becomes

$$y^2 \phi''(y) + 3y\phi'(y) + \lambda^2 y^2 \phi(y) = 0$$

$$\phi(A) = \phi(\hat{B}) = 0, \qquad \text{where } \hat{B} = A + Ba,$$

and where $\mu = -(\lambda B)^2$. In terms of y this is a regular Sturm–Liouville problem with countably many orthogonal eigenfunctions in $L_2(A, \hat{B}, y^3)$. The eigenfunction equation can be solved by matching it with (3.10). We find that the general solution is

$$\phi(y) = \frac{1}{y} \left[c_1 J_1(\lambda y) + c_2 Y_1(\lambda y) \right].$$

We obtain a nontrivial solution satisfying the boundary conditions provided the determinant of the linear system

$$\begin{pmatrix} J_1(\lambda A) & Y_1(\lambda A) \\ J_1(\lambda \hat{B}) & Y_1(\lambda \hat{B}) \end{pmatrix} \begin{pmatrix} c_1 \\ c_2 \end{pmatrix} = 0$$

is zero. Hence the eigenvalues are determined by the nonzero roots of the equation

$$f(\lambda) \equiv J_1(\lambda A)Y_1(\lambda \hat{B}) - J_1(\lambda \hat{B})Y_1(\lambda A) = 0.$$

For each root λ_n we obtain the eigenfunction

$$\phi_n(y) = \frac{1}{Y} \left[Y_1(\lambda_n A)J_1(\lambda_n y) - J_1(\lambda_n A)Y_1(\lambda_n y) \right].$$

The general theory assures us that the eigenfunctions are linearly independent and that in the $L_2(A, \hat{B}, y^3)$ sense they approximate the source term

$$ch'/h^3 = cB/y^3.$$

In terms of y and z the approximating problem is solved by

$$p_N(y, z) = \sum_{n=1}^{N} \alpha_n(z)\phi_n(y)$$

where $\alpha_n(z)$ is a solution of

$$-(\lambda_n B)^2 \alpha_n(z) + \alpha_n''(z) = \hat{\gamma}_n = \frac{\langle cB/y^3, \phi_n \rangle}{\langle \phi_n, \phi_n \rangle}$$

$$\alpha_n(0) = \alpha_n(b) = 0,$$

with $\langle f, g \rangle = \int_A^{\hat{B}} f(y)g(y)y^3 dy$. Its solution is

$$\alpha_n(z) = \frac{\hat{\gamma}_n}{\lambda_n^2} \left[\frac{\sinh B\lambda_n(b - z) + \sinh B\lambda_n z}{\sinh B\lambda_n b} - 1 \right].$$

The roots of $f(\lambda) = 0$ and the $\hat{\gamma}_n$ require numerical computation. (We point out that the singular Sturm–Liouville problem with an assumed film thickness

$$h(x) = Bx$$

has no eigenfunctions on $[0, a]$ because $Y_1(y) \to -\infty$ and $J_1(y)/y \to 1/2$ as $y \to 0$ so that $\phi(0) = 0$ requires $c_1 = c_2 = 0$. For this case an eigenfunction expansion in terms of eigenfunctions in z is proposed in [17] but it is not clear that the resulting solution can satisfy both boundary conditions

$$p(0, z) = p(a, z) = 0.)$$

Example 8.6 Lubrication of a step bearing.

The following interface problem yields the fluid pressures u and v inside a step bearing [17]

$$\Delta u = 0, \qquad (x, z) \in D_1 = (0, x_0) \times (0, b)$$

$$\Delta v = 0, \qquad (x, z) \in D_2 = (x_0, a) \times (0, b).$$

The pressures are ambient outside the slider, i.e.

$$u(0, z) = v(b, z) = 0$$
$$u(x, 0) = u(x, b) = 0, \qquad v(x, 0) = v(x, b) = 0.$$

The interface conditions representing continuity of pressure and fluid film flow at the interface at x_0 are

$$u(x_0, z) = v(x_0, z)$$
$$u_x(x_0, z) = c_1 v_x(x_0, z) + c_2$$

for $c_1 > 0$.

It follows immediately that the approximate solutions u_N and v_N can be written as

$$u_N(x, z) = \sum_{n=1}^{N} \alpha_n(x)\phi_n(z)$$

$$v_N(x, z) = \sum_{n=1}^{N} \beta_n(x)\phi_n(z)$$

where

$$\phi_n(z) = \sin \lambda_n z, \qquad \lambda_n = \frac{n\pi}{b}$$

and

$$\alpha_n(x) = \hat{\alpha}_n \sinh \lambda_n x$$
$$\beta_n(x) = \hat{\beta}_n \sinh \lambda_n(x - a).$$

The interface conditions require

$$\hat{\alpha}_n \sinh \lambda_n x_0 = \hat{\beta}_n \sinh \lambda_n(x_0 - a)$$

$$\hat{\alpha}_n \lambda_n \cosh \lambda_n x_0 = c_1 \hat{\beta}_n \lambda_n \cosh \lambda_n(x_0 - a) + c_2.$$

After a little algebra we obtain

$$\hat{\alpha}_n = \frac{c_2 \sinh \lambda_n(x_0 - a)}{\lambda_n [\cosh \lambda_n x_0 \sinh \lambda_n(x_0 - a) - c_1 \sinh \lambda_n x_0 \cosh \lambda_n(x_0 - a)]}$$

$$\hat{\beta}_n = \hat{\alpha}_n \frac{\sinh \lambda_n x_0}{\sinh \lambda_n(x_0 - a)}.$$

Example 8.7 The Dirichlet problem on an L-shaped domain.

It has become commonplace to solve elliptic boundary value problems numerically with the technique of domain decomposition to reduce the geometric complexity of the computational domain, and to allow parallel computations. The ideas of domain decomposition are equally relevant for the analytic eigenfunction solution method. There are a variety of decomposition methods on overlapping and nonoverlapping domains. Here we shall look briefly at combining eigenfunction expansions with the classical Schwarz alternating procedure [3].

Let us summarize the Schwarz method for the Dirichlet problem

$$\Delta u = F \quad \text{in } D$$

$$u = 0 \quad \text{on } \partial D$$

where D is an L-shaped domain formed by the union of the rectangles $D_1 = (0, a) \times (0, b)$ and $D_2 = (0, c) \times (0, d)$ for $c < a$ and $b < c$. To simplify the exposition we shall choose here

$$D_1 = (0, 2) \times (0, 1), \qquad D_2 = (0, 1) \times (0, 2)$$

and

$$F(x, y) \equiv 1.$$

The Schwarz alternating procedure generates a solution u in $D = D_1 \cup D_2$ as the limit of sequences

$$u = \lim_{k \to \infty} \{u^k, v^k\}$$

where u^k and v^k are the solutions for $k = 1, 2, \ldots$ of the Dirichlet problems

$$\Delta u^k = 1, \qquad (x, y) \in D_1$$

$$u^k(0, y) = u^k(2, y) = u^k(x, 0) = 0$$

$$u^k(x, 1) = \begin{cases} v^{k-1}(x, 1), & x \in (0, 1) \\ 0, & x \in [1, 2] \end{cases}$$

and

$$\Delta v^k = 1, \qquad (x, y) \in D_2$$

$$v^k(0, y) = v^k(x, 2) = v^k(x, 0) = 0$$

$$v^k(1, y) = \begin{cases} u^k(1, y), & y \in (0, 1) \\ 0, & y \in [1, 2] \end{cases}$$

for any continuous initial guess $v^0(x, 1)$, $x \in [0, 1]$ with $v^0(1, 1) = 0$.

Let $\{\phi_m(x)\}$ denote the eigenfunctions associated with Δu and $\{\psi_m(y)\}$ those corresponding to Δv. For this simple geometry we see that

$$\phi_m(x) = \sin \lambda_m x$$

$$\psi_m(y) = \sin \lambda_m y$$

with $\lambda_m = \frac{m\pi}{2}$. Then the projected source term is

$$P_M 1 = \sum_{m=1}^{M} \hat{\gamma}_m \phi_m(x) = \sum_{m=1}^{M} \hat{\gamma}_m \psi_m(y)$$

where

$$\hat{\gamma}_m = \frac{\langle 1, \phi_m \rangle}{\langle \phi_m, \phi_m \rangle} \quad \text{with } \langle f, g \rangle = \int_0^2 f(x)g(x)dx.$$

The eigenfunction solutions of

$$\Delta u^k = P_M 1$$

and

$$\Delta v^k = P_M 1$$

are

$$u_M^k(x, y) = \sum_{m=1}^{M} \alpha_m^k(y)\phi_m(x)$$

$$v_M^k(x, y) = \sum_{m=1}^{M} \beta_m^k(x)\psi_m(y),$$

where $\alpha_m^k(y)$ is the solution of

$$-\lambda_m^2 \alpha_m + \alpha_m''(y) = \hat{\gamma}_m$$

$$\alpha_m(0) = 0, \qquad \alpha_m(1) = \hat{\alpha}_m^k$$

with

$$\hat{\alpha}_m^k = \frac{\langle \langle v^{k-1}(x, 1), \phi_m(x) \rangle \rangle}{\langle \phi_m, \phi_m \rangle}$$

$\langle\langle f, g \rangle\rangle$ is meant to denote the integral

$$\int_0^1 f(x)g(x)dx$$

that arises because $v(x, 1) \equiv 0$ for $x \in [1, 2]$. Similarly, $\beta_m^k(x)$ is the solution of

$$-\lambda_m^2 \beta_m + \beta_m''(x) = \hat{\gamma}_m$$

$$\beta_m(0) = 0, \qquad \beta_m(1) = \hat{\beta}_m^k$$

with

$$\hat{\beta}_m^k = \frac{\langle\langle u^k(1, y), \psi_m(y) \rangle\rangle}{\langle \psi_m, \psi_m \rangle}$$

We verify by inspection that

$$\alpha_m^k(y) = \hat{\alpha}_m^k f_m(y) + p_m(y)$$

where

$$f_m(y) = \frac{\sinh \lambda_m y}{\sinh \lambda_m}$$

$$p_m(y) = \frac{\hat{\gamma}_m}{\lambda_m^2} \left[\frac{\sinh \lambda_m(1 - y) + \sinh \lambda_m y}{\sinh \lambda_m} - 1 \right].$$

Similarly, $\beta_m^k(x)$ will have the form

$$\beta_m^k(x) = \hat{\beta}_m^k g_m(x) + q_m(x)$$

for appropriate g and q. (Of course, for this geometry and source term we know that $g_m = f_m$ and $q_m = p_m$.) It follows that

$$\hat{\alpha}_m^k = \sum_{j=1}^M \frac{\langle\langle (g_j(x)\hat{\beta}_j^{k-1} + q_j(x))\psi_j(1), \phi_m(x) \rangle\rangle}{\langle \phi_m, \phi_m \rangle} \qquad (8.13)$$

With $\vec{\alpha} = (\hat{\alpha}_1, \ldots, \hat{\alpha}_M)$ and $\vec{\beta} = (\hat{\beta}_1, \ldots, \hat{\beta}_M)$ we can rewrite (8.13) in matrix form

$$\vec{\alpha}^k = \mathcal{A}\vec{\beta}^{k-1} + \vec{b}$$

where

$$\mathcal{A}_{mj} = \frac{\langle\langle g_j(x)\psi_j(1), \phi_m(x) \rangle\rangle}{\langle \phi_m, \phi_m \rangle}$$

$$b_m = \sum_{j=1}^M \frac{\langle\langle q_j(x)\psi_j(1), \phi_m(x) \rangle\rangle}{\langle \phi_m, \phi_m \rangle}.$$

Similarly, we obtain

$$\vec{\beta}^k = \mathcal{B}\vec{\alpha}^k + \vec{c}$$

where

$$\mathcal{B}_{mj} = \frac{\langle\langle f_j(y)\phi_j(1), \psi_m(y)\rangle\rangle}{\langle\psi_m, \psi_m\rangle}$$

$$c_m = \sum_{j=1}^{M} \frac{\langle\langle p_j(y)\phi_j(1), \psi_m(y)\rangle\rangle}{\langle\psi_m, \psi_m\rangle}.$$

These matrix equations lead to the recursion formula

$$\vec{\alpha}^k = \mathcal{A}\mathcal{B}\vec{\alpha}^{k-1} + \mathcal{A}\vec{c} + \vec{b}. \tag{8.14}$$

Since u_M^k and v_M^k are analytic solutions, we know from the Schwarz alternating principle that they converge to a solution u_M of

$$\Delta u = P_M 1.$$

Hence

$$u_M(x,y) = \sum_{j=1}^{M} \left[f_j(y)\alpha_j^* + p_j(y) \right] \phi_j(x), \qquad (x,y) \in D_1$$

$$v_M(x,y) = \sum_{j=1}^{M} \left[g_j(x)\beta_j^* + q_j(x) \right] \psi_j(y), \qquad (x,y) \in D_2$$

where $\vec{\alpha}^*$ and $\vec{\beta}^*$ are solutions of the algebraic equations

$$\vec{\alpha}^* = \mathcal{A}\mathcal{B}\vec{\alpha}^* + \mathcal{A}\vec{c} + \vec{b}, \qquad \vec{\beta}^* = \mathcal{B}\vec{\alpha}^* + \vec{c}. \tag{8.15}$$

On $D_1 \cap D_2$ both formulas yield the same function.

Fig. 8.7 shows the solution of our problem obtained by solving the linear system (8.15) and substituting into u_M and v_M. The symmetry in this problem simplifies our calculations but is not essential for this approach. It will be applicable to the union of intersecting domains provided that the Schwarz alternating principle applies for the geometries in question, and that separation of variables is applicable on each subdomain.

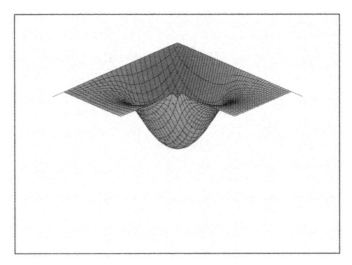

Figure 8.7: Surface u_{15} for $\Delta u = 1$ in D, $u = 0$ on ∂D for an L-shaped domain D.

Example 8.8 Poisson's equation in polar coordinates.

We consider the problem

$$\Delta u = F \text{ on } D$$

where D is

i) A truncated wedge with $D = \{(r,\theta) : 0 \leq R_0 < r < R_1, \theta_0 < \theta < \theta_1\}$.

ii) An annulus with $D = \{(r,\theta) : 0 \leq R_0 < r < R_1, \theta_0 < \theta \leq \theta_0 + 2\pi\}$. If $R_0 = 0$, we have a disk of radius R_1.

iii) An exterior domain with $D = \{(r,\theta) : 0 < R_0 < r, \theta_0 < \theta < \theta_1\}$.

For ease of notation we shall choose our coordinate system such that $\theta_0 = 0$. The boundary of D consists of

i) The rays $\theta = \theta_0, \theta_1$ and the arcs $r = R_0, R_1$ between these rays. If $R_0 = 0$, then the arc $r = R_0$ shrinks to the origin.

ii) The circles $r = R_0, R_1$. If $R_0 = 0$, then the inner circle becomes the origin.

iii) The rays $\theta = \theta_0, \theta_1$ and the arc $r = R_0$. There is no boundary at infinity, but for computational reasons it is common to think of an outer boundary at $r = R_1$ where $R_1 \gg R_0$.

We shall consider first the problem on a wedge

$$\mathcal{L}u \equiv u_{rr} + \frac{1}{r} u_r + \frac{1}{r^2} u_{\theta\theta} = F(r,\theta), \quad (r,\theta) \in D = (R_0, R_1) \times (\theta_0, \theta_1)$$

$$c_1(r,\theta)\frac{\partial u}{\partial \theta} + c_2(r,\theta)u = g(r,\theta), \qquad \theta = \theta_1, \theta_2, \qquad (8.16)$$

$$c_1(r,\theta)\frac{\partial u}{\partial n} + c_2(r,\theta)u = g(r,\theta), \qquad r = R_0, R_1$$

where $c_1(r,\theta)$ and $c_2(R,\theta)$ are *constants* along the rays and arcs. Note that these boundary conditions are not the same as

$$c_1 \frac{\partial u}{\partial n} + c_2 u = g(r,\theta)$$

because

$$\frac{\partial u}{\partial n} = \frac{1}{r} u_\theta(r,\theta) \text{ on } \theta = \theta_1$$

which introduces an r-dependence into the coefficient of u_θ.

If the functions $g(r,\theta_0)$ and $g(r,\theta_1)$ are twice continuously differentiable with respect to r, then as in Example 8.2 we can find a smooth function

$$v(r,\theta) = g(r,\theta_0)f_1(\theta) + g(r,\theta_1)f_2(\theta)$$

which satisfies the boundary conditions on the two rays. Then the function $w(r,\theta) = u(r,\theta) - v(r,\theta)$ satisfies homogeneous boundary conditions on the rays. Hence with little loss of generality we shall assume that

$$g(r,\theta_0) = g(r,\theta_1) = 0.$$

With this simplification the eigenfunction method for Poisson's equation in polar coordinates differs little from that for Poisson's equation on a rectangle described in the preceding examples. We consider

$$\Delta u = F(r,\theta)$$

$$c_1 u_\theta + c_2 u = 0 \text{ on } \theta = \theta_0, \theta_1 \qquad (8.17)$$

$$c_1 \frac{\partial u}{\partial n} + c_2 u = g(r,\theta) \text{ on } r = R_0, R_1.$$

The eigenvalue problem associated with (8.17) is

$$\phi''(\theta) = \mu\phi(\theta)$$

$$c_1(r,\theta_0)\phi'(\theta_0) + c_2(r,\theta_0)\phi(\theta_0) = 0$$

$$c_1(r, \theta_1)\phi'(\theta_1) + c_2(r, \theta_1)\phi(\theta_1) = 0.$$

This is a standard problem with solutions $\{\phi_n(\theta)\}$ either given in Table 3.1 or found after solving equation (3.6).

The approximating problem for (8.17) is

$$r^2 u_{rr} + r u_r + u_{\theta\theta} = P_N(r^2 F(r, \theta)) = \sum_{n=n_0}^{N} \gamma_n(r)\phi_n(\theta)$$

$$-c_1(R_0, \theta)u_r + c_2(R_0, \theta)u = P_N g(R_0, \theta) = \sum_{n=n_0}^{N} \hat{a}_n \phi_n(\theta)$$

$$c_1(R_1, \theta)u_r + c_2(R_1, \theta)u = P_N g(R_1, \theta) = \sum_{n=n_0}^{N} \hat{\delta}_n \phi_n(\theta)$$

where $n_0 = 0$ or 1 depending on the boundary conditions, and where

$$\gamma_n(r) = \frac{r^2 \langle F(r, \theta), \phi_n(\theta) \rangle}{\langle \phi_n, \phi_n \rangle} \quad \text{with} \quad \langle f, g \rangle = \int_{\theta_0}^{\theta_1} f(\theta)g(\theta)d\theta.$$

The solution is

$$u_N(r, \theta) = \sum_{n=n_0}^{N} \alpha_n(r)\phi_n(\theta)$$

where

$$r^2 \alpha_n'' + r \alpha_n' - \lambda_n^2 \alpha_n = \gamma_n(r) \tag{8.18}$$

$$-c_1(R_0, \theta)\alpha_n'(R_0) + c_2(R_0, \theta)\alpha_n(R_0) = \hat{a}_n$$

$$c_1(R_1, \theta)\alpha_n'(R_1) + c_2(R_1, \theta)\alpha_n(R_1) = \hat{\delta}_n.$$

Equation (8.18) is an inhomogeneous Cauchy–Euler equation and has the solution

$$\alpha_n(r) = d_{1n} + d_{2n} \ln r + \alpha_{np}(r) \qquad \text{if } \lambda_n = 0$$
$$\alpha_n(r) = d_{1n} r^{\lambda_n} + d_{2n} r^{-\lambda_n} + \alpha_{np}(r) \quad \text{if } \lambda_n > 0.$$

The coefficients d_{1n} and d_{2n} are determined from the boundary conditions, while the particular integral $\alpha_{np}(r)$ can often be computed with the method of undetermined coefficients or the method of variation of parameters.

Let us now comment on the case either where $R_0 = 0$ or where the exterior problem has to be solved. For a realistic problem on a wedge with $R_0 = 0$ we do not have an inner boundary condition but would expect that

$$\lim_{r \to 0} |u(r, \theta)| < \infty.$$

In this case necessarily $d_{2n} = 0$ for all n.

For the exterior problem the application typically imposes a decay condition on u as $r \to \infty$. For example, if $\lim u(r, \theta) \to 0$ as $r \to \infty$, then necessarily

$$d_{1n} = 0 \text{ for all } n \geq 0 \text{ as well as } d_{2n} = 0 \text{ when } \lambda_n = 0.$$

If the boundary data do not allow us to find a smooth function v such that $w = u - v$ satisfies a homogeneous boundary condition on the two rays then we are forced into a formal splitting

$$u = u_1 + u_2$$

where u_1 solves (8.17) and u_2 is a solution of

$$u_{rr} + \frac{1}{r} u_r + \frac{1}{r^2} u_{\theta\theta} = 0$$

$$c_1(r, \theta) \frac{\partial u}{\partial n} + c_2(r, \theta) u = 0 \quad \text{for } r = R_0, R_1,$$

$$c_1(r, \theta) u_\theta + c_2(r, \theta) u = g(r, \theta) \quad \text{for } \theta = \theta_0, \theta_1.$$

The associated eigenvalue problem is

$$r^2 \phi''(r) + r \phi'(r) = \mu \phi(r)$$

$$-c_1(R_0, \theta) \phi'(R_0) + c_2(R_0, \theta) \phi(R_0) = 0$$

$$c_1(R_1, \theta) \phi'(R_1) + c_2(R_1, \theta) \phi(R_1) = 0.$$

The discussion of Section 3.2 implies that for $R_0 > 0$ the above eigenvalue problem can be written as a regular Sturm–Liouville problem for the equation

$$(r\phi'(r))' = \mu \frac{1}{r} \phi(r).$$

Hence we know that the eigenvalues are nonpositive, i.e., $\mu = -\lambda^2$, and that the eigenfunctions are orthogonal with respect to the weight function

$$w(r) = \frac{1}{r}.$$

The Cauchy–Euler form of the differential equation allows us to find the eigenfunctions explicitly as

$$\phi(r) = d_1 \cos \lambda \ln r + d_2 \sin \lambda \ln r$$

(see Example 6.9). Substitution into the boundary conditions leads to the matrix equation

$$A(\lambda) \begin{pmatrix} d_1 \\ d_2 \end{pmatrix} = 0$$

where

$$A_{11} = c_1(R_0)\frac{\lambda}{R_0}\sin(\lambda \ln R_0) + c_2(R_0)\cos(\lambda \ln R_0)$$

$$A_{12} = -c_1(R_0)\frac{\lambda}{R_0}\cos(\lambda \ln R_0) + c_2(R_0)\sin(\lambda \ln R_0)$$

$$A_{21} = c_1(R_1)\frac{\lambda}{R_1}\sin(\lambda \ln R_1) + c_2(R_1)\cos(\lambda \ln R_1)$$

$$A_{22} = -c_1(R_1)\frac{\lambda}{R_1}\cos(\lambda \ln R_1) + c_2(R_1)\sin(\lambda \ln R_1).$$

As in our discussion of Chapter 3, we now require that $\det A = 0$. This is in general a nonlinear equation in λ. For each solution λ_n we find the corresponding eigenfunction

$$\phi_n(r) = A_{12}(\lambda_n)\cos(\lambda_n \ln r) - A_{11}(\lambda_n)\sin(\lambda_n \ln r).$$

Once the eigenvalues and eigenfunctions are known we can write the approximating problem and solve it in the span of the eigenfunctions. Details are now very problem dependent and will not be pursued further here.

Let us now turn to periodic solutions. If the problem is given on an annulus or a disk and the solution is expected to be periodic in θ, then we may take

$$\theta_0 = 0 \text{ and } \theta_1 = 2\pi$$

and impose the homogeneous periodicity conditions

$$u(r, 0) = u(r, 2\pi)$$

$$u_\theta(r, 0) = u_\theta(r, 2\pi).$$

The eigenvalue problem for $\phi(\theta)$ is now

$$\phi''(\theta) = \mu\phi(\theta)$$

$$\phi(0) = \phi(2\pi)$$

$$\phi'(0) = \phi'(2\pi).$$

The eigenvalues are

$$\mu_n = -\lambda_n^2 = -n^2, \qquad n = 0, 1, 2, \ldots.$$

For $n = 0$ we have the eigenfunction

$$\psi_0(\theta) = 1.$$

For $n > 0$ we have two linearly independent eigenfunctions

$$\phi_n(\theta) = \sin n\theta$$

$$\psi_n(\theta) = \cos n\theta.$$

For any square integrable function f defined on $[0, 2\pi]$ we find that

$$P_N f(\theta) = a_0 \psi_0(\theta) + \sum_{n=1}^{N} [a_n \phi_n(\theta) + b_n \psi_n(\theta)]$$

is just the Nth partial sum of its Fourier series.

For a given N one could linearly order the $2N + 1$ eigenfunctions

$$\{\phi_1, \phi_2, \ldots, \phi_N, \psi_0, \ldots, \psi_N\}$$

and use the notation of (8.18) for the eigenfunction solution of Poisson's equation. However, it is more convenient to write

$$u_N(r, \theta) = \alpha_0(r) + \sum_{n=1}^{N} [\alpha_n(r)\phi_n(\theta) + \beta_n(r)\psi_n(\theta)] \qquad (8.19)$$

where both $\alpha_n(r)$ and $\beta_n(r)$ satisfy equation (8.18) and the boundary conditions obtained by projecting $g(R_0, \theta)$ and $g(R_1, \theta)$ into span$\{\phi_n\}\cup$span$\{\psi_n\}$. It follows that

$$\begin{aligned}
\alpha_0(r) &= d_{10} + d_{20} \ln r + \alpha_{0p}(r) \\
\alpha_n(r) &= d_{1n} r^n + d_{2n} r^{-n} + \alpha_{np}(r) \qquad (8.20) \\
\beta_n(r) &= D_{1n} r^n + D_{2n} r^{-n} + \alpha_{np}(r).
\end{aligned}$$

If instead of an annulus with $R_0 > 0$ we have a disk, then as above we expect that

$$\lim_{r \to 0} |u(r, \theta)| < \infty$$

so that necessarily

$$d_{2n} = 0 \quad \text{and} \quad D_{2n} = 0 \quad \text{for all } n.$$

Similarly, decay at infinity in the exterior problem would demand

$$d_{20} = d_{1n} = D_{1n} = 0 \quad \text{for all } n.$$

Example 8.9 Steady-state heat flow around an insulated pipe I.

We consider the following thermal problem for temperatures u and v:

$$\Delta u = 0 \qquad R_0 < r < R_1$$

$$\Delta v = 0 \qquad R_1 < r < R_2$$

$$u(R_0) = 80, \qquad v(R_2) = -10$$

with interface conditions of continuity of temperature and heat flux

$$u(R_1) = v(R_1)$$

$$\alpha u_r(R_1) = v_r(R_1), \qquad \alpha < 1.$$

The equations describe, for example, radial heat flow around a vertical pipe with an insulation layer of thickness $R_1 - R_0$ whose conductivity is α times the conductivity k of the material in the annulus $R_1 < r < R_2$. The aim is to find R_1 such that

$$u(R_1) = 0.$$

(This would give us an estimate, for example, of how much insulation is needed to keep an insulated oil production pipe in permafrost from melting the surrounding soil.)

For ease of calculation we shall assume that the variable r has been scaled so that $R_0 = 1$. Since there is no angular dependence, we know from (8.19) and (8.20) that

$$u(r) = 80 + d_{20} \ln r$$

$$v(r) = D_{10} + D_{20} \ln r.$$

The boundary, interface, and target condition $u(R_1) = 0$ lead to the following algebraic system:

$$D_{10} + D_{20} \ln R_2 = -10$$

$$80 + d_{20} \ln R_1 = D_{10} + D_{20} \ln R_1$$

$$\alpha \frac{d_{20}}{R_1} = \frac{D_{20}}{R_1}$$

$$80 + d_{20} \ln R_1 = 0.$$

Since $D_{20} = \alpha d_{20}$, we can write these equations in matrix form as

$$\begin{pmatrix} 1 & \alpha \ln R_2 \\ -1 & (1-\alpha) \ln R_1 \\ 0 & \ln R_1 \end{pmatrix} \begin{pmatrix} D_{10} \\ d_{20} \end{pmatrix} = \begin{pmatrix} -10 \\ -80 \\ -80 \end{pmatrix}.$$

Gaussian elimination shows that this system has a unique solution if and only if

$$\frac{(1-\alpha)\ln R_1 + \alpha \ln R_2}{\ln R_1} = +\frac{9}{8}$$

so that

$$R_1 = R_2^{\frac{8\alpha}{8\alpha+1}}.$$

We note that the insulation layer $(R_1 - 1) \to 0$ as the conductivity $\alpha k \to 0$ and that $R_1 \to R_2$ as $\alpha \to \infty$, which is the correct thermal limiting behavior.

Example 8.10 Steady-state heat flow around an insulated pipe II.

Let us now tackle the analogous problem for a buried pipeline in a soil with prescribed linear temperature profile. We shall assume that the center of the pipe of radius R_0 is the origin which lies at a depth of $50R_0$, that the annulus $R_0 < r < R_1$ is filled with insulation, and that R_2 is the radius of the region around the pipe heated by it. The temperature of the pipe is 80 degrees, and the temperature in the soil for $r > R_2$ increases with depth according to

$$T(y) = -10 + \beta(50R_0 - y)$$

where β is a known parameter. The aim is to find R_1 such that the maximum temperature is zero on the outer edge of the insulation, whose conductivity is again α times the conductivity in the annulus $R_1 < r < R_2$.

The model equations are

$$\Delta u = 0 \qquad R_0 < r < R_1$$

$$\Delta v = 0 \qquad R_1 < r < R_2$$

$$u(R_0, \theta) = 80$$

$$u(R_1, \theta) = v(R_1, \theta)$$

$$\alpha u_r(R_1, \theta) = v_r(R_1, \theta)$$

$$v(R_2, \theta) = T(R_2 \sin \theta).$$

Now there is angular dependence. It follows from the above discussion that

$$u(r, \theta) = \alpha_0(r) + \sum_{n=1}^{N} [\alpha_n(r)\phi_n(\theta) + \beta_n(r)\psi_n(\theta)]$$

$$v(r, \theta) = A_0(r) + \sum_{n=1}^{N} [A_n(r)\phi_n(\theta) + B_n(r)\psi_n(\theta)].$$

The boundary condition $u(R_0, \theta) = 80$ implies that

$$\alpha_0(R_0) = 80$$

and

$$\alpha_n(R_0) = \beta_n(R_0) = 0 \qquad \text{for } n \geq 1.$$

The interface conditions imply that for all n

$$\alpha_n(R_1) = A_n(R_1)$$
$$\alpha[\alpha_n'(R_1)] = A_n'(R_1)$$
$$\beta_n(R_1) = B_n(R_1)$$
$$\alpha[\beta_n'(R_1)] = B_n'(R_1).$$

The boundary condition $v(R_2, \theta) = -10 + 50\beta R_0 - \beta R_2 \sin\theta$ implies that

$$A_0(R_2) = -10 + 50\beta R_0$$
$$A_1(R_2) = -\beta R_2$$
$$A_n(R_2) = 0 \qquad \text{for } n \geq 2$$
$$B_n(R_2) = 0 \qquad \text{for } n \geq 1.$$

It follows by inspection that

$$\alpha_n(r) = A_n(r) = 0 \qquad \text{for } n \geq 2$$

and

$$\beta_n(r) = B_n(r) = 0 \qquad \text{for } n \geq 1.$$

Hence the problem reduces to determining the coefficients of

$$\alpha_0(r) = d_{10} + d_{20} \ln r$$
$$A_0(r) = D_{10} + D_{20} \ln r$$
$$\alpha_1(r) = d_{11}r^1 + d_{12}r^{-1}$$
$$A_1(r) = D_{11}r^1 + D_{12}r^{-1}$$

so that the boundary and interface conditions are satisfied.

This requires the solution of the linear system

$$
\begin{pmatrix}
1 & \ln R_0 & 0 & 0 & 0 & 0 & 0 & 0 \\
0 & 0 & 0 & 0 & R_0 & \frac{1}{R_0} & 0 & 0 \\
1 & \ln R_1 & -1 & -\ln R_1 & 0 & 0 & 0 & 0 \\
0 & 0 & 0 & 0 & R_1 & \frac{1}{R_1} & -R_1 & -\frac{1}{R_1} \\
0 & \frac{1}{R_1} & 0 & -\frac{1}{R_1} & 0 & 0 & 0 & 0 \\
0 & 0 & 0 & 0 & \alpha & -\frac{\alpha}{R_1^2} & -1 & \frac{1}{R_1^2} \\
0 & 0 & 1 & \ln R_2 & 0 & 0 & 0 & 0 \\
0 & 0 & 0 & 0 & 0 & 0 & R_2 & \frac{1}{R_2}
\end{pmatrix}
\begin{pmatrix}
d_{10} \\ d_{20} \\ D_{10} \\ D_{20} \\ d_{11} \\ d_{12} \\ D_{11} \\ D_{12}
\end{pmatrix}
=
\begin{pmatrix}
80 \\ 0 \\ 0 \\ 0 \\ 0 \\ 0 \\ -10+50\beta R_0 \\ -\beta R_2
\end{pmatrix}.
$$

Since the temperature in the soil increases monotonically with depth, the warmest point on $r = R_1$ will be directly below the center of the pipe so that $\theta = \frac{3\pi}{2}$. Given the fixed parameters R_0, R_2, and β it now is a simple matter to search for $R_1 \in (R_0, R_2)$ such that

$$
u\left(R_1, \frac{3\pi}{2}\right) = d_{10} + d_{20}\ln R_1 + (d_{11}R_1 + d_{12}R_1^{-1})\sin\left(\frac{3\pi}{2}\right) = 0.
$$

Example 8.11 Poisson's equation on a triangle.

When the equations arising in eigenfunction expansions are sufficiently simple, it becomes possible to determine boundary data on rectangular boundaries which approximate prescribed boundary data on curved boundaries. Let us illustrate the process with the following model problem:

$$
\begin{aligned}
\Delta u &= 1 && \text{in } T \\
u &= 0 && \text{on } \partial T
\end{aligned}
$$

where T is the triangle with vertices $A = (0,0)$, $B = (3,0)$, $C = (2,1)$.

We shall imbed T into the circular wedge $D = \{(r,\theta) : 0 < r < R\} = 3$, $0 < \theta < \theta_1 = \tan^{-1}.5$ and solve

$$
\begin{aligned}
\Delta u &= 1 && \text{in } D \\
u &= 0 && \text{on } \theta = 0, \theta_1 \\
u &= f(\theta) && \text{on } r = R
\end{aligned}
$$

where f is a yet unknown function to be determined such that on the line segment $\overline{BC}$

$$u(r, \theta) \cong 0.$$

The associated eigenvalue problem

$$\phi'' = -\lambda^2 \phi$$
$$\phi(0) = \phi(\theta_1) = 0$$

has the solution

$$\phi_n(\theta) = \sin \lambda_n \theta \quad \text{where } \lambda_n = \frac{n\pi}{\theta_1}.$$

If we write

$$P_N f(\theta) = \sum_{n=1}^{N} \hat{\beta}_n \phi_n(\theta)$$

$$P_N 1 = \sum_{n=1}^{N} \hat{\gamma}_n \phi_n(\theta),$$

then the problem

$$\Delta u = P_N 1 \qquad \text{in } D$$
$$u = 0 \qquad \text{on } \theta = 0, \theta_1$$
$$u = P_N f \qquad \text{on } r = R$$

has the eigenfunction expansion solution

$$u_N(r, \theta) = \sum_{n=1}^{N} \alpha_n(r) \phi_n(\theta)$$

where

$$r^2 \alpha_n'' + r\alpha_n' - \lambda_n^2 \alpha_n = r^2 \hat{\gamma}_n$$
$$\alpha_n(0) = 0, \qquad \alpha_n(R) = \hat{\beta}_n.$$

The solution is

$$\alpha_n(r) = \hat{\beta}_n \left(\frac{r}{R}\right)^{\lambda_n} + \frac{\hat{\gamma}_n}{4 - \lambda_n^2} \left(r^2 - R^2 \left(\frac{r}{R}\right)^{\lambda_n}\right).$$

It is in general not possible to choose the N parameters $\{\hat{\beta}_n\}$ such that $u_N \equiv 0$ along the line segment $\overline{BC}$, i.e., along $r(\theta) = \frac{3}{\sin\theta + \cos\theta}$. However, we can find $\{\hat{\beta}_n\}$ such that

$$E\left(\hat{\beta}_1, \ldots, \hat{\beta}_n\right) = \int_0^{\theta_1} u_N^2(r(\theta), \theta) d\theta$$

is minimized. Necessary and sufficient for this problem is that $\vec{\beta}$ be a solution of the equations

$$\frac{\partial E}{\partial \hat{\beta}_i} = 0, \qquad i = 1, \ldots, N.$$

A little algebra shows that the vector $\vec{\beta} = \{\hat{\beta}_1, \ldots, \hat{\beta}_N\}$ must satisfy the linear system

$$\mathcal{A}\vec{\beta} = b$$

where

$$\mathcal{A}_{ij} = \left\langle \left(\frac{r(\theta)}{R}\right)^{\lambda_j} \phi_j(\theta), \left(\frac{r(\theta)}{R}\right)^{\lambda_i} \phi_i(\theta) \right\rangle$$

$$b_i = \sum_{n=1}^{N} \frac{\hat{\gamma}_n}{4 - \lambda_n^2} \left\langle \left(r(\theta)^2 - R^2 \left(\frac{r(\theta)}{R}\right)^{\lambda_n}\right) \phi_n(\theta), \left(\frac{r(\theta)}{R}\right)^{\lambda_i} \phi_i(\theta) \right\rangle.$$

These inner products have to be evaluated numerically.

Alternatively, one could compute $\{\hat{\beta}_n\}$ by collocation such that

$$u_N(r(\theta_n), \theta_n) = 0$$

for N distinct values of θ. The choice of collocation points is critical for success of this method and requires familiarity with the theory of collocation. The least squares method, in contrast, appears to be automatic and quite robust. The inner products involve very smooth trigonometric functions and are easily evaluated numerically. We show in Fig. 8.11 a plot of $u_N(r, \theta)$ for $N = 10$.

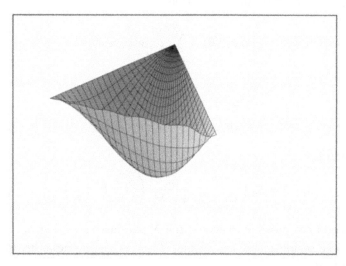

Figure 8.11: Surface u_{10} for $\Delta u = 1$ in T, $u = 0$ on ∂T for a triangle T.

We do not have a proof that u_N converges to the solution u of the original problem on the triangle T as $N \to \infty$, although numerical experiments suggest that it does so. The question of convergence, however, is not relevant in this case. We know that u_N is an analytic solution of Poisson's equation

$$\Delta u = P_N 1,$$

and we can observe the calculated solution $u_N(r(\theta), \theta)$ on ∂D. For the surface $u_{10}(r, \theta)$ shown in Fig. 8.11 we find

$$\max_\theta |u_{10}(r(\theta), \theta)| = 4 \ 10^{-3}.$$

As will be shown in Section 8.3, this is enough a posteriori information to judge by elementary means whether u_N is a useful approximation of u.

8.2 Eigenvalue problem for the two-dimensional Laplacian

When we turn to diffusion and vibration problems involving two spacial variables, we shall need the eigenfunctions of the Laplacian in the plane. As mentioned in Chapter 3 we know that the general eigenvalue problem

$$\nabla \cdot p \nabla u = \mu u \qquad \text{in } D, \text{ with } p > 0$$

$$c_1(x, y) \frac{\partial u}{\partial n} + c_2(x, y) u = 0 \qquad \text{on } \partial D$$

with

$$c_1 c_2 \geq 0, \qquad c_1^2 + c_2^2 > 0$$

has countably many nonpositive eigenvalues $\{\mu_n\}$ and eigenfunctions $\{\Phi_n\}$ which are orthogonal in $L_2(D)$. However, an explicit computation of eigenvalues and eigenfunctions can be carried out only for very special problems. Foremost among these is the eigenvalue problem for the Laplacian in orthogonal coordinates which is just a special case of a potential problem in the plane.

Example 8.12 The eigenvalue problem for the Laplacian on a rectangle.

We shall consider the simplest problem

$$\Delta u = \mu u, \qquad (x, y) \in D = (0, a) \times (0, b)$$

$$u = 0 \text{ on } \partial D.$$

We associate with the Laplacian and the boundary condition the familiar eigenvalue problem

$$\phi''(x) = -\lambda^2 \phi(x)$$

$$\phi(0) = \phi(a) = 0$$

which has the solutions

$$\phi_m(x) = \sin \lambda_m x, \qquad \lambda_m = \frac{m\pi}{a}.$$

A solution of the eigenvalue problem in the subspace span$\{\phi_m\}$ would have to be of the form

$$u_M(x,y) = \sum_{m=1}^{M} \alpha_m(y)\phi_m(x) \tag{8.21}$$

where

$$-\lambda_m^2 \alpha_m + \alpha_m'' = \mu \alpha_m$$

$$\alpha_m(0) = \alpha_m(b) = 0.$$

However, for a given λ_m this is an eigenvalue problem for α_m. It has a nontrivial solution $\alpha_{mn}(y) = \sin \rho_n y$ only if

$$\rho_n^2 \equiv \mu + \lambda_m^2 = -\left(\frac{n\pi}{b}\right)^2 \qquad \text{for an integer } n > 0.$$

Conversely, if we set

$$\mu_{mn} = -\left[\lambda_m^2 + \left(\frac{n\pi}{b}\right)^2\right],$$

then

$$u_{mn}(x,y) = \alpha_{mn}(y)\phi_m(x)$$

is an eigenfunction with eigenvalue μ_{mn}. Since for $k \neq n$ we have $\mu_{mk} \neq \mu_{mn}$, no linear combination of $\{u_{mn}\}$ can be an eigenfunction. Hence each eigenfunction expansion (8.21) for given m can consist of only one term like $u_{mn}(x,y)$, but there are countably many different expansions because $n = 1, 2, \ldots$. Thus for the Laplacian on the square we obtain the eigenfunctions

$$u_{mn}(x,y) = \sin \lambda_m x \sin \rho_n y$$

with corresponding eigenvalues

$$\mu_{mn} = -[\lambda_m^2 + \rho_n^2] = -\left[\left(\frac{m\pi}{a}\right)^2 + \left(\frac{n\pi}{b}\right)^2\right], \qquad m, n = 1, 2 \ldots.$$

Moreover

$$\int_D u_{k\ell}(x,y)u_{mn}(x,y)dx\,dy = \begin{cases} 0 & \text{if } m \neq k \text{ or } \ell \neq n \\ ab/4 & \text{if } m = k \text{ and } n = \ell. \end{cases}$$

Note that is is possible that

$$\mu_{mn} = \mu_{k\ell}$$

for distinct indices; however, the corresponding eigenfunctions remain orthogonal.

Example 8.13 The Green's function for the Laplacian on a square.

The availability of eigenfunctions for the Laplacian on a domain D suggests an alternate, and formally at least, simpler method for solving Poisson's equation

$$\Delta u = F(x,y) \text{ in } D$$
$$u = 0 \qquad \text{on } \partial D.$$

If $\{\mu_n, \Phi_n\}$ are eigenvalues and eigenfunctions of

$$\Delta \Phi(x,y) = \mu \Phi(x,y) \text{ in } D$$
$$\Phi(x,y) = 0 \qquad \text{on } \partial D,$$

then we compute the projection

$$F_N(x,y) = \sum_{n=1}^{N} \gamma_n \Phi_n(x,y)$$

where

$$\gamma_n = \frac{\langle F, \Phi_n \rangle}{\langle \Phi_n, \Phi_n \rangle}$$

with

$$\langle f, g \rangle = \int_D fg \, dx \, dy.$$

We observe that

$$\Delta u = F_N(x,y)$$

is solved exactly by

$$u_N(x,y) = \sum_{n=1}^{N} \alpha_n \Phi_n(x,y)$$

when

$$\alpha_n = \frac{\gamma_n}{\mu_n},$$

since $\mu_n \neq 0$ for all n.

Let us illustrate this approach, and contrast it to the one-dimensional eigenfunction expansion used earlier, by computing the Green's function $G(x,y,\xi,\eta)$ for the Laplacian on a rectangle.

The problem is stated as follows. Find a function $G(x, y, \xi, \eta)$ which as a function of x and y satisfies (formally)

$$\Delta G(x, y, \xi, \eta) = -\delta(x - \xi)\delta(y - \eta), \qquad (x, y) \in D = (0, a) \times (0, b)$$
$$G(x, y, \xi, \eta) = 0, \qquad\qquad\qquad (x, y) \in \partial D$$

where (ξ, η) is an arbitrary but fixed point in D. Here δ denotes the so-called delta (or impulse) function. We shall avoid the technical complications inherent in a rigorous definition of the delta function by thinking of it as the pointwise limit as $k \to \infty$ of the function

$$\delta_k(x) = \begin{cases} k^2(x + 1/k), & x \in [-1/k, 0] \\ k^2(1/k - x), & x \in (0, 1/k] \\ 0 & \text{otherwise.} \end{cases}$$

We note that for any function f which is continuous at $x = 0$ we obtain the essential feature of the delta function

$$\int_I f(x)\delta(x)dx = \lim_{k \to \infty} \int_I f(x)\delta_k(x)dx = f(0)$$

where I is any open interval containing $x = 0$.

The eigenfunctions of the Laplacian for these boundary conditions were computed above as

$$u_{mn}(x, y) = \sin \lambda_m x \sin \rho_n y$$

with corresponding eigenvalues

$$\mu_{mn} = -\left[\lambda_m^2 + \rho_n^2\right]$$

where

$$\lambda_m = \frac{m\pi}{a}, \qquad \rho_n = \frac{n\pi}{b}.$$

The (formal) projection of $-\delta(x - \xi)\delta(y - \eta)$ onto the span of the eigenfunctions is

$$P_{MN}(-\delta)(x, y) = \sum_{\substack{m=1 \\ n=1}}^{M,N} \gamma_{mn} u_{mn}(x, y)$$

where

$$\gamma_{mn} = \frac{\langle -\delta(x - \xi)\delta(y - \eta), u_{mn}\rangle}{\langle u_{mn}, u_{mn}\rangle} = \frac{-4 \sin \lambda_m \xi \sin \rho_n \eta}{ab}.$$

Hence $G_{MN}(x, y, \xi, \eta) = \sum_{m,n=1}^{M,N} \alpha_{mn} u_{mn}(x, y)$ is the solution of the approximating problem provided

$$\alpha_{mn} = \frac{\gamma_{mn}}{\mu_{mn}}.$$

Thus

$$G_{MN}(x, y, \xi, \eta) = \frac{4}{ab} \sum_{m,n=1}^{M,N} \frac{\sin \lambda_m \xi \sin \lambda_m x \sin \rho_n \eta \sin \rho_n y}{\lambda_m^2 + \rho_n^2}$$

is our approximation to the Green's function.

The alternative is to solve the problem

$$\Delta u = P_M(-\delta(x - \xi)\delta(y - \eta)) \qquad \text{in } D$$
$$u = 0 \qquad\qquad\qquad \text{on } D$$

with an eigenfunction expansion in terms of $\phi_m(x) = \sin \lambda_m x$, $\lambda_m = m\pi/a$. Then

$$P_M(-\delta(x - \xi)\delta(y - \eta)) = \sum_{m=1}^{M} \gamma_m \sin \lambda_m x$$

with (formally)

$$\gamma_m = \frac{-\langle \delta(x - \xi)\delta(y - \eta), \phi_m \rangle}{\langle \phi_m, \phi_m \rangle} = \frac{-2}{a} \delta(y - \eta) \sin \lambda_m \xi \equiv \hat{\gamma}_m \delta(y - \eta).$$

The solution of the approximating problem is given by

$$u_M(x, y) = \sum_{m=1}^{M} \alpha_m(y) \sin \lambda_m x$$

where $\alpha_m(y)$ solves

$$-\lambda_m^2 \alpha_m + \alpha_m''(y) = \hat{\gamma}_m \delta(y - \eta)$$
$$\alpha_m(0) = \alpha_m(b).$$

The solution of this problem is known to be the scaled one-dimensional Green's function $-\hat{\gamma}_m \alpha_m(y, \eta)$ where

$$\alpha_m(y, \eta) = \begin{cases} \frac{\sinh \lambda_m \eta \sinh \lambda_m (b-y)}{\lambda_m \sinh \lambda_m b} & 0 < \eta < y \\ \frac{\sinh \lambda_m y \sinh \lambda_m (b-\eta)}{\lambda_m \sinh \lambda_m b} & y < \eta < b. \end{cases}$$

The approximation to the Green's function is now

$$G_M(x, y, \xi, \eta) = \sum_{m=1}^{M} \frac{2}{a} \sin \lambda_m \xi \sin \lambda_m x \alpha_m(y, \eta).$$

The two approximations $G_{MN}(x, y, \xi, \eta)$ and $G_M(x, y, \xi, \eta)$ are different functions. In fact, it can be shown that $G_{MN}(x, y, \xi, \eta) = P_N G_M(x, y, \xi, \eta)$ where $P_N G_M$ denotes the projection of G_M onto span $\{\sin \rho_n y\}$.

The analytic Green's function is of the form

$$G(x, y, \xi, \eta) = s(x, y, \xi, \eta) + \phi(x, y, \xi, \eta)$$

where s is the fundamental solution of Laplace's equation in R_2

$$s(x, y, \xi, \eta) = -\frac{1}{2\pi} \ln \|(x, y) - (\xi, \eta)\|$$

and where ϕ is a smooth solution of the Dirichlet problem

$$\Delta\phi = 0, \qquad\qquad \text{for } (x, y) \in D$$
$$\phi = -s(x, y, \xi, \eta), \qquad \text{for } (x, y) \in \partial D.$$

The Dirchlet problem for ϕ is, of course, almost a model problem for an eigenfunction expansion solution. Note that the boundary function is not defined and smooth on D because of the singularity at $(x, y) = (\xi, \eta)$. s is smooth on ∂D and one could subtract the boundary values on opposite sides to zero out the boundary data at, say $x = 0$ and $x = 1$, and solve the problem. However, for $(\xi, \eta) = (.5, .5)$ the symmetry of the problem suggests preconditioning the problem with (8.4) and solving the new problem with a formal splitting. This way no new source terms arise, and the two problems of the splitting are symmetric in x and y. The eigenfunction solution ϕ_{20} is very accurate so that the computed G is accepted as the analytic Green's function $G(x, y, .5, .5)$.

Fig. 8.13 shows a plot of the three approximate Green's functions $G_{10\ 10}(x, y, .5, .5)$, $G_{100}(x, y, .5, .5)$, and $G(x, y, .5, .5)$ along the diagonal $x = y$ of D. The agreement between G and G_N appears very good, particularly in view of the common use of Green's functions under an integral. A plot of the Green's function approximation G_{MN} for $M = N = 100$ (not shown) yields a Green's function approximation indistinguishable from G_{100}.

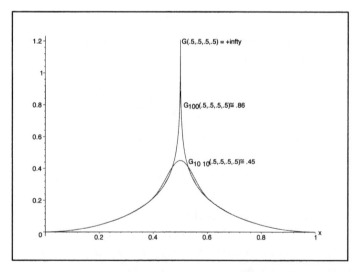

Figure 8.13: Green's functions G and approximations $G_{10\ 10}$, G_{100} for the Laplacian on a square. Shown are $G(x, x, .5, .5)$ and its approximations.

Example 8.14 The eigenvalue problem for the Laplacian on a disk.

We wish to find the eigenvalues and eigenvectors of

$$\Delta u(r, \theta) = -\delta^2 u(r, \theta) \qquad \text{on } 0 \le r < R$$
$$u(R, \theta) = 0,$$

where $\mu = -\delta^2$. The problem in polar coordinates is

$$r^2 u_{rr} + r u_r + u_{\theta\theta} = -\delta^2 r^2 u(r, \theta) \tag{8.22}$$
$$\lim_{r \to 0} |u(r, \theta)| < \infty, \qquad u(R, \theta) = 0.$$

The associated one-dimensional eigenvalue problem is again the periodic problem in θ given by

$$\phi''(\theta) = -\lambda^2 \phi(\theta)$$
$$\phi(0) = \phi(2\pi)$$
$$\phi'(0) = \phi'(2\pi).$$

It has the eigenvalues $-\lambda_n^2 = -n^2$ for $n = 0, 1, \ldots$. For $n = 0$ the eigenfunction is

$$\psi_0(\theta) = 1.$$

For all other eigenvalues we obtain the two eigenfunctions

$$\psi_n(\theta) = \cos n\theta$$

$$\phi_n(\theta) = \sin n\theta.$$

For each eigenfunction $\phi_n(\theta)$ and $\psi_n(\theta)$ we shall find all eigenfunctions of the Laplacian of the form

$$u^1(r, \theta) = \alpha(r)\phi_n(\theta)$$
$$u^2(r, \theta) = \beta(r)\psi_n(\theta).$$

Substitution into the eigenvalue equation (8.22) shows that $\alpha(r)$ and $\beta(r)$ must satisfy

$$r^2 \alpha''(r) + r\alpha'(r) - n^2 \alpha(r) = -\delta^2 r^2 \alpha(r), \qquad n = 1, 2, \ldots$$
$$r^2 \beta''(r) + r\beta'(r) - n^2 \beta(r) = -\delta^2 r^2 \beta(r), \qquad n = 0, 1, 2, \ldots$$
$$\lim_{r \to 0} |\alpha(r)| < \infty, \qquad\qquad \lim_{r \to 0} |\beta(r)| < \infty,$$

and $\alpha(R) = \beta(R) = 0$. For a given integer n this equation is known to be Bessel's equation. Let us pick an arbitrary integer n. Then we have the two general solutions

$$\alpha(r) = c_{n1}J_n(\delta r) + c_{n2}Y_n(\delta r)$$

$$\beta(r) = d_{n1}J_n(\delta r) + d_{n2}Y_n(\delta r).$$

Finiteness at $r = 0$ requires that $c_{n2} = d_{n2} = 0$ and since eigenfunctions are only determined up to a multiplicative constant, we shall set $c_{n1} = d_{n1} = 1$. $J_n(\delta R) = 0$ then demands that δR be a root of the Bessel function $J_n(x)$. Let x_{mn} denote the mth nonzero root of the Bessel function $J_n(x)$. We write

$$\delta_{mn} = \frac{x_{mn}}{R}$$

and obtain the solutions

$$\alpha_m(r) = \beta_m(r) = J_n(\delta_{mn}r).$$

It follows that the Laplacian on a disk has countably many eigenvalues

$$\mu_{mn} = -\delta_{mn}^2 = -\left(\frac{x_{mn}}{R}\right)^2, \qquad m = 1, 2, \ldots, \quad n = 0, 1, \ldots$$

and the associated eigenfunctions

$$u_{m0}^2(r, \theta) = J_0(\delta_{m0}r) \qquad \text{for } n = 0$$

and

$$\begin{aligned} u_{mn}^1(r, \theta) &= J_n(\delta_{mn}r)\sin n\theta \\ u_{mn}^2(r, \theta) &= J_n(\delta_{mn}r)\cos n\theta \end{aligned} \qquad \text{for } n = 1, 2, \ldots.$$

Example 8.15 The eigenvalue problem for the Laplacian on the surface of a sphere.

We want to find the eigenvalues and eigenfunctions of

$$\Delta u = -\delta^2 u \qquad \text{on } D$$

where D is the surface of a sphere of radius R.

It is natural to center the sphere at the origin and express the surface points in spherical coordinates

$$x = R\cos\theta\sin\phi$$

$$y = R\sin\theta\sin\phi$$

$$z = R\cos\phi$$

for $\theta \in [0, 2\pi]$ and $\phi \in [0, \pi]$. Then

$$\Delta u = -\delta^2 u \quad \text{on } r = R$$

becomes

$$\frac{1}{R^2 \sin^2 \phi} \frac{\partial u^2}{\partial \theta^2} + \frac{1}{R^2 \sin \phi} \frac{\partial}{\partial \phi} \left(\sin \phi \frac{\partial u}{\partial \phi} \right) = -\delta^2 u.$$

We can rewrite this equation as

$$\Delta_{(\theta, \phi)} u \equiv \frac{\partial^2 u}{\partial \theta^2} + \sin \phi \frac{\partial}{\partial \phi} \left(\sin \phi \frac{\partial u}{\partial \phi} \right) = -\delta^2 \sin^2 \phi \, u$$

where for convenience we have set $R = 1$.

The eigenfunctions are 2π periodic in θ and need to remain finite at the two poles of the sphere where $\phi = 0$ and $\phi = \pi$. The associated one-dimensional eigenvalue problem is

$$\Phi''(\theta) = -\lambda^2 \Phi(\theta)$$

$$\Phi(\theta) = \Phi(2\pi), \qquad \Phi'(\theta) = \Phi'(2\pi).$$

In general there are two linearly independent eigenfunctions

$$\Phi_n(\theta) = \sin n\theta, \qquad n = 1, 2, \ldots$$

$$\Psi_n(\theta) = \cos n\theta, \qquad n = 0, 1, 2, \ldots$$

corresponding to the eigenvalue $-\lambda_n^2 = -n^2$. Again, we shall look for all eigenfunctions of the Laplacian of the form

$$u^1(\theta, \phi) = \alpha(\phi) \Phi_n(\theta)$$

$$u^2(\theta, \phi) = \beta(\phi) \Psi_n(\theta).$$

Substitution into the eigenvalue equation shows that α and β must satisfy

$$-n^2 \alpha + \sin \phi \frac{d}{d\phi} \left(\sin \phi \frac{d\alpha}{d\phi} \right) = -\delta^2 \sin^2 \phi \alpha$$

$$-n^2 \beta + \sin \phi \frac{d}{d\phi} \left(\sin \phi \frac{d\beta}{d\phi} \right) = -\delta^2 \sin^2 \phi \beta.$$

This equation is well known in the theory of special functions where it is usually rewritten in terms of the variable $z = \cos \phi$. An application of the chain rule leads to the equivalent equation

$$(1 - z^2) \frac{d^2 \alpha}{dz^2} - 2z \frac{d\alpha}{dz} + \left(\delta^2 - \frac{n^2}{1 - z^2} \right) \alpha = 0.$$

The same equation holds for $\beta(z)$. This equation is called the "associated Legendre equation." It is known [1] that it has solutions which remain finite at the poles $z = \pm 1$ if and only if

$$\delta^2 = m(m+1) \qquad \text{for } m \geq 0$$

in which case the solution is the so-called "associated Legendre function of the first kind" $P_m^n(z)$. For the special case of $n = 0$ the corresponding associated Legendre function $P_m^0(z)$ is usually written as $P_m(z)$ and is known as the mth order Legendre polynomial. The Legendre polynomials customarily are scaled so that $P_m(1) = 1$. For $m \leq 3$ they are

$$P_0(z) = 1, \quad P_1(z) = z, \quad P_2(z) = \frac{1}{2}(3z^2 - 1), \quad P_3(z) = \frac{1}{2}(5z^3 - 3z).$$

The associated Legendre functions of the first kind are found from

$$P_m^n(z) = (1 - z^2)^{n/2} \frac{d^n}{dz^n} P_m(z)$$

or alternatively, from

$$P_m^n(z) = \frac{1}{2^m m!} (1 - z^2)^{n/2} \frac{d^{m+n}}{dz^{m+n}} (z^2 - 1)^m.$$

This expression shows that

$$P_m^n(z) \equiv 0 \qquad \text{for } n > m.$$

(We remark that the second fundamental solution of the associated Legendre equation is the associated Legendre function of the second kind. However, it blows up logarithmically at the poles, i.e., at $z = \pm 1$, and will not be needed here.) Returning to our eigenvalue problem we see that for each $n \geq 0$ there are countably many eigenvalues and eigenfunctions

$$-\delta_{mn}^2 = -m(m+1), \qquad m = n, n+1 \ldots$$

$$u_{mn}^1(\theta, \phi) = P_m^n(\cos \phi)\Phi_n(\theta), \qquad n = 1, 2, \ldots$$

$$u_{mn}^2(\theta, \phi) = P_m^n(\cos \phi)\Psi_n(\theta), \qquad n = 0, 1, \ldots.$$

These eigenfunctions are mutually orthogonal on the surface of the sphere so that

$$\langle u_{mn}^i, u_{kl}^j \rangle = 0$$

unless $m = k$, $n = \ell$, and $i = j$. Here

$$\langle f(\theta, \phi), g(\theta, \phi) \rangle = \int_0^{2\pi} \int_0^\pi f(\theta, \phi) g(\theta, \phi) \sin \phi \, d\phi \, d\theta.$$

Since the one-dimensional eigenfunctions $\{\Phi_n(\theta), \Psi_n(\theta)\}$ are orthogonal in $L_2(0, 2\pi)$, it follows that $\{P_m^n(\cos\phi)\}$ are orthogonal in $L_2(0, \pi, \sin\phi)$, i.e., that

$$\int_0^{\pi} P_m^n(\cos\phi)P_k^{\ell}(\cos\phi)\sin\phi\, d\phi = \int_{-1}^{1} P_m^n(z)P_k^{\ell}(z)dz$$

$$= 0 \qquad \text{for } m \neq k \text{ or } n \neq \ell.$$

(Since $P_m(z)$ is an mth order polynomial, this implies that $\{P_m(z)\}_{m=0}^{N}$ is an orthogonal basis in $L_2(-1,1)$ of the subspace of Nth order polynomials.) Moreover, it can be shown that

$$\int_{-1}^{1} P_m^n(z)^2 dz = \frac{2(m+n)!}{(m-n)!(2m+1)}.$$

8.3 Convergence of $u_N(x,y)$ to the analytic solution

In Chapter 1 we pointed out that the maximum principle can provide an error bound on the pointwise error $u(x,y) - u_N(x,y)$ in terms of the approximation errors for the source term and the boundary data. An application of these ideas is presented in Chapter 6 for the diffusion equation. We shall revisit these issues by examining the error incurred in solving Poisson's equation on a triangle as described in Example 8.11.

We recall the problem: the solution u of the problem

$$\Delta u = 1 \qquad (x,y) \in T$$
$$u = 0 \qquad (x,y) \in \partial T$$

where T is a triangle with vertices $(0,0),(3,0),(2,1)$, is approximated with the solution u_N of the problem

$$\Delta u = P_N 1$$

$$u = 0 \qquad \text{on } \overline{(0,0)(3,0)} \quad \text{and} \quad \overline{(0,0)(2,1)},$$

i.e., $u = 0$ on $\theta = 0, \theta_1$, where $\theta_1 = \tan^{-1}.5$, and

$$u = u_N(r(\theta),\theta) \qquad \text{on } \overline{(3,0)(2,1)},$$

where $u_N(r(\theta),\theta)$ is obtained from the least squares minimization. $P_N 1$ is the projection in $L_2(0,\theta_1)$ of $f(\theta) \equiv 1$ into span $\{\sin\lambda_n\theta\}$.

Since u_N is an exact solution of Poisson's equation in T, we can write

$$u_N(x,y) = u_{1N}(x,y) + u_{2N}(x,y)$$

where

$$\Delta u_{1N} = P_N 1 \qquad \text{in } T$$
$$u_{1N} = 0 \qquad \text{on } \partial T,$$
$$\Delta u_{2N} = 0 \qquad \text{in } T$$
$$u_{2N} = u_N(x,y) \qquad \text{on } \partial T.$$

Our goal is to estimate the error

$$u - u_N = (u - u_{1N}) - u_{2N}.$$

As a solution of Laplace's equation u_{2N} must assume its maximum and minimum on ∂T. Hence

$$|u_{2N}(x,y)| \le \max_\theta |u_N(r(\theta),\theta)|.$$

However

$$\max_{(x,y)\in \bar{T}} |1 - P_N 1| \ge 1 \qquad \text{for all } N$$

because $P_n 1 = 0$ for $\theta = 0$. The failure of $P_N 1$ to converge uniformly precludes an application of the maximum principle to estimate $(u - u_{1N})$.

We know from the Sturm–Liouville theorem (and Chapter 4) that $P_N 1$ converges to $f(\theta) \equiv 1$ in the $\mathcal{L}_2(0,\theta_1)$ norm. This implies the inequality

$$\|P_N 1 - 1\|_{\mathcal{L}_2(T)} \le \sqrt{4.5}\|P_N 1 - 1\|_{\mathcal{L}_2(0,\theta_1)}$$

because $r(\theta) \le 3$.

We employ now a so-called energy estimate. Let us write

$$w = u - u_{1N}.$$

Then w is an analytic solution of

$$\Delta w = 1 - P_N 1 \qquad \text{in } T$$
$$w = 0 \qquad \text{on } \partial T.$$

It follows from the divergence theorem and the fact that $w = 0$ on ∂T that

$$\int_T w\Delta w\, dx\, dy = \int_T [\nabla \cdot w\nabla w - \nabla w \cdot \nabla w]\, dx\, dy$$
$$= \oint_{\partial T} w \frac{\partial w}{\partial n}\, ds - \int_T \nabla w \cdot \nabla w\, dx\, dy,$$

i.e., that

$$\int_T \nabla w \cdot \nabla w\, dx\, dy = -\int_T [1 - P_N 1]\, w\, dx\, dy. \qquad (8.23)$$

For any point $(x, y) \in T$ we can write

$$|w(x,y)|^2 = \left| w(x,0) + \int_0^y \frac{\partial w}{\partial \eta}(x,\eta)d\eta \right|^2 \le y \int_0^y \left(\frac{\partial w}{\partial \eta} \right)^2 d\eta$$

where the last inequality comes from Schwarz's inequality applied to $\int_0^y 1 \frac{\partial w}{\partial \eta}(x,\eta)dy$. Since $y \le 1$ for all $(x,y) \in T$, integration over T yields the Poincaré inequality

$$\int_T w^2 dx\, dy \le \int_T \nabla w \cdot \nabla w dx\, dy. \qquad (8.24)$$

If we also apply Schwarz's inequality to the right side of (8.23), we obtain from (8.23), (8.24) the estimate

$$\|w\|_{\mathcal{L}_2(T)} \le \|1 - P_N 1\|_{\mathcal{L}_2(T)}.$$

The pointwise bound on $|u_{2N}|$ implies

$$\|u_{2N}\|_{\mathcal{L}_2(T)} < \sqrt{\text{area}(T)} \max_\theta |u_{2N}(r(\theta),\theta)|.$$

The actual approximation error is then bounded by

$$\|u - u_N\|_{\mathcal{L}_2(T)} \le \sqrt{1.5} \max_\theta |u_{2N}(r(\theta),\theta)| + \sqrt{4.5}\,\|1 - P_N 1\|_{\mathcal{L}_2(0,\theta_1)}.$$

For the computed solution of Example 8.14 we observe for $N = 15$

$$\max_\theta |u_N(r(\theta),\theta)| \cong .0031$$

and a simple calculation shows that

$$\|P_{15}1 - 1\|_{\mathcal{L}_2(0,\theta)} \cong .1083.$$

Obviously, $N = 15$ is much too small to give a tight bound on the actual error (see also the discussion on p. 56). The need for many terms of the Fourier expansion to overcome the Gibbs phenomenon has arisen time and again and applies here as well, but the computed solutions u_N do not appear to change noticeably for larger N.

We shall end our comments on convergence by pointing out that the error estimating techniques familiar to finite element practitioners also apply to eigenfunction expansions. To see this we shall denote by u the solution of the model problem

$$\Delta u = F \qquad \text{in } D = (0,a) \times (0,b)$$
$$u = 0 \qquad \text{on } \partial D.$$

The approximate solution is

$$u_N(x,y) = \sum_{n=1}^{N} \alpha_n(y)\phi_n(x)$$

where $\phi_n(x) = \sin \lambda_n x$, and where u_N solves

$$\Delta u_N = P_N F \qquad \text{in } D$$
$$u_N = 0 \qquad \text{on } \partial D.$$

Let $v_k(x,y) = \beta(y)\phi_k(x)$ where $\beta(y)$ is any smooth function such that $\beta(0) = \beta(b) = 0$. Let $\langle\ ,\ \rangle$ and $\|\ \|$ denote inner product and norm in $\mathcal{L}_2(D)$ and let us set

$$A(u,v) \equiv \langle \nabla u, \nabla v \rangle \equiv \int_D \nabla u \cdot \nabla v dx\, dy.$$

Then

$$\langle v_k, \Delta(u - u_N) \rangle = -\langle A(v_k, u - u_N) \rangle = \langle v_k, F - F_N \rangle = 0 \qquad (8.25)$$

because

$$\langle v_k, F - F_N \rangle = \int_0^b \int_0^a \beta(y)\phi_k(x) \left[F(x,y) - \sum_{n=1}^{N} \gamma_n(y)\phi_n(x) \right] dx\, dy = 0$$

due to the orthogonality of $\{\phi_n\}$ in $\mathcal{L}_2(0,a)$. Because A is bilinear, we see that

$$A(u - u_N, u - u_N) = A(u - v_N + v_N - u_N, u - u_N) = A(u - v_N, u - u_N) \quad (8.26)$$

for any function v_N of the form

$$v_N(x,y) = \sum_{n=1}^{N} \beta_n(y)\phi_n(x), \qquad \beta_n(0) = \beta_n(b) = 0. \qquad (8.27)$$

Schwarz's inequality applied to (8.26) yields

$$\|\nabla(u - u_N)\| \leq \|\nabla(u - v_N)\|.$$

We conclude that the error in the gradient is bounded by the error in the gradient of the best possible approximation to u of the form (8.27). The error analysis now has become a question of approximation theory. This view of error analysis is developed in the finite element theory where very precise $\mathcal{L}_2$-estimates for the error and its gradient are based on bounds for the analytic solution u and its derivatives known from the theory of partial differential equations. The consequence for the eigenfunction expansion method is that the error in our approximation is no larger than the error which arises when the (unknown) solution $u(x, y)$ for the above Dirichlet problem is expanded in a two-dimensional Fourier sine series

$$P_N u(x, y) = \sum_{\substack{m=1 \\ n=1}}^{M,N} \alpha_{mn} \sin \frac{m\pi}{b} y \sin \frac{n\pi}{a} x$$

because this is an expression of the form (8.27).

Exercises

8.1) Solve

$$\Delta u = 0$$
$$u(0, y) = \sqrt{(y - 1/2)^2} + y/10$$
$$u(1, y) = 2y/5$$
$$u(x, 0) = (1 - x)/2$$
$$u(x, 1) = \sqrt{(x - 3/5)^2} \, .$$

8.2) Consider the problem

$$\Delta u = 0 \qquad \text{in } D = (0, 1) \times (0, 1)$$

$$u = g(x, y) = \sqrt{x} \, y \qquad \text{on } \partial D.$$

i) Find an eigenfunction approximation of u applied to a formal splitting without preconditioning.

ii) Find an eigenfunction approximation of u applied to a formal splitting after preconditioning.

iii) Find an eigenfunction approximation of u after zeroing out the data on $x = 0$ and $x = 1$.

iv) Find an eigenfunction approximation of u after zeroing out the data on $y = 0$ and $y = 1$.

Discuss which of the above results appears most acceptable for finding the solution u of the original problem.

8.3) In Example 8.6 find a piecewise linear function $f(x)$ such that

$$f(0) = f(b) = 0$$
$$f(x_0-) = f(x_0+)$$
$$f'(x_0-) = c_1 f'(x_0+) + c_2.$$

Find the interface problem satisfied by

$$w_1 = u - f, \qquad w_2 = v - f$$

and solve it in terms of eigenfunctions $\{\phi_n(x)\}$ discussed in Section 3.3.

8.4) Find the solution of

$$\Delta \phi = 0, \quad (x, y) \in D = (0, 1) \times (0, 1)$$
$$\phi = \frac{1}{2\pi} \ln \|(x, y) - (.95, .95)\|, \qquad (x, y) \in \partial D.$$

Plot the approximate Green's functions of Example 8.13 for $(\xi, \eta) = (.95, .95)$ and various $M = N = 10, 20, 100$ as well as the analytic Green's function

$$G(x, y, .95, .95) = -\frac{1}{2\pi} \ln \|(x, y) - (.95, .95)\| + \phi_{20}(x, y)$$

over the square minus a disk of radius $\epsilon \ll 1$ centered at $(.95, .95)$. Comment on the quality of the approximations.

8.5) Let the boundary condition

$$c_1(x, y)\frac{\partial u}{\partial n} + c_2(x, y)u = g(x, y)$$

hold on the boundary of the rectangle $D = (0, a) \times (0, b)$. Assume that c_1 and c_2 are constant along the sides of D, and that u, u_x, u_y, and u_{xy} are continuous on ∂D at (a, b). Show that the two boundary equations and their first derivatives are consistent at (a, b) if and only if condition (8.8) holds.

8.6) State and prove an analogue of Theorem 6.13 for the problem

$$\Delta w = F \quad \text{in } D = (0, a) \times (0, b)$$
$$w = 0 \quad \text{on } \partial D.$$

8.7) Find an eigenfunction solution of the biharmonic problem

$$\Delta^2 u = \Delta(\Delta u) = 1, \qquad (x, y) \in D = (0, 1) \times (0, 1)$$

$$u = \frac{\partial^2 u}{\partial n^2} = 0 \quad \text{on } \partial D$$

when it is written in the form

$$\Delta u = v, \qquad (x, y) \in D$$

$$\Delta v = 1, \qquad (x, y) \in D$$

$$u(x, y) = 0, \qquad (x, y) \in \partial D$$

$$v(x, y) = 0, \qquad (x, y) \in \partial D$$

(cf. Example 7.9).

8.8) Write out in detail the steps required for an eigenfunction solution of the problem

$$\Delta u = F(r, \theta), \qquad 0 < R_0 < r < R_1, \quad 0 < \theta < 2\pi$$

$$u(R_0, \theta) = f(\theta), \qquad u(R_1, \theta) = g(\theta)$$

$$u(r, 0) = u(r, 2\pi) + h(r)$$

$$u_\theta(r, 0) = u_\theta(r, 2\pi) + k(r).$$

Chapter 9

Multidimensional Problems

The algorithm of Chapter 5 requires linearly independent, preferably orthogonal, eigenfunctions but otherwise is independent of the dimension of the eigenvalue problem. Employing the eigenfunctions for the Laplacian in the plane we shall solve in broad outline some representative diffusion, vibration, and potential problems involving two- and three-space dimensions.

9.1 Applications of the eigenfunction expansion method

Example 9.1 A diffusive pulse test.

We consider the inverse problem of determining a diffusion constant A such that the solution $u(x, y, t)$ of

$$A\Delta u - u_t = -100\delta(x - 1)\delta(y - 1)\delta(t - 1), \quad (x, y) \in D = (0, 3) \times (0, 4), \quad t > 0$$

$$u = 0 \qquad \text{on } \partial D$$

$$u(x, y, 0) = 0 \qquad \text{in } D$$

at the point $(x, y) = (2, 3)$ assumes its maximum at time $t = 2$.

As in Example 8.13 the calculations are formal but can be made rigorous when the δ-functions are thought of as the limit of piecewise linear continuous functions.

We know from Example 8.12 that the eigenvalue problem

$$\Delta\Phi = \mu\Phi \quad \text{in } D$$

$$\Phi = 0 \qquad \text{on } \partial D$$

has the eigenfunctions

$$u_{mn}(x,y) = \sin \lambda_m x \sin \rho_n y, \qquad \lambda_m = \frac{m\pi}{3}, \qquad \rho_n = \frac{n\pi}{4}$$

with corresponding eigenvalues

$$\mu_{mn} = -[\lambda_m^2 + \rho_n^2].$$

The approximating problem is

$$A\Delta u - u_t = -100\delta(t-1)P_{MN}\delta(x-1)\delta(y-1)$$

$$= \delta(t-1) \sum_{m,n=1}^{M,N} \hat{\gamma}_{mn} u_{mn}(x,y)$$

where

$$\hat{\gamma}_{mn} = \frac{-100\langle \delta(x-1)\delta(y-1), u_{mn}\rangle}{\langle u_{mn}, u_{mn}\rangle} = \frac{-100\sin\lambda_m \sin\rho_n}{3}.$$

The problem is solved by

$$u_{MN}(x,u,t) = \sum_{m,n=1}^{M,N} \alpha_{mn}(t) u_{mn}(x,y)$$

where

$$A\mu_{mn}\alpha_{mn}(t) - \alpha'_{mn}(t) = \delta(t-1)\hat{\gamma}_{mn}$$

$$\alpha_{mn}(0) = 0.$$

It follows from the variation of parameters solution that

$$\alpha_{mn}(t) = \begin{cases} 0 & t < 1 \\ -\hat{\gamma}_{mn}\exp(A\mu_{mn}(t-1)), & t > 1. \end{cases}$$

Hence we need to determine A such that

$$u_{MN}(2,3,t) = \sum_{m,n=1}^{M,N} -\hat{\gamma}_{mn}\exp(A\mu_{mn}(t-1))\sin\lambda_m 2\sin\rho_n 3$$

assumes its maximum at $t = 2$. For a given A the function $u_{MN}(2,3,t)$ is readily evaluated so that the value $t_{\max}(A)$ can be found where it achieves its maximum. Since $t_{\max}(A)$ is monotonely decreasing with increasing A, it then is simple to search for A^* such that $t_{\max}(A^*) = 2$. A search over $(t, A) \in [1, 2.5] \times [.5, 1]$ with $dt = .01$ and $dA = .005$ yields the results

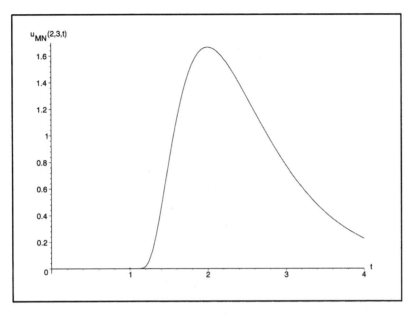

Figure 9.1: Pulse $u_{MN}(2,3,t)$ for the diffusion constant $A^* = .775$ found with a bisection method. $M = N = 90$.

$M = N$	A^*
12	.775
50	.775
90	.775

A plot of the pulse $u_{MN}(2,3,t)$ for $A^* = .775$ is shown in Fig. 9.1 for $M = N = 90$. While A^* appears to be computable from relatively few terms, the pulse itself requires many more terms to adequately approximate the δ-functions.

For a problem with different homogeneous boundary data new eigenfunctions for the Laplacian are needed, but their computation along the lines of Example 8.12 is straightforward. A more involved problem arises when the boundary data are not homogeneous. Consider, for example, the general Dirichlet problem

$$\mathcal{L}u \equiv \Delta u - u_t = F(x,y,t), \qquad (x,y) \in D = (0,a) \times (0,b), \quad t > 0$$

$$u = g(x,y,t) \quad \text{on } \partial D, \quad t > 0$$
$$u(x,y,0) = u_0(x,y) \quad \text{in } D.$$

If g is smooth in D for $t > 0$, then rewriting the problem for $w = u - g$ yields a new Dirichlet problem with zero boundary data. If g is only smooth on ∂D, then we can define

$$v(x, y, t) = \left[g(0, y, t)\frac{a - x}{a} + g(a, y, t)\frac{x}{a} \right]$$

$$+ \left[g(x, 0, t) - g(0, 0, t)\frac{a - x}{a} - g(a, 0, t)\frac{x}{a} \right] \frac{b - y}{b}$$

$$+ \left[g(x, b, t) - g(0, b, t)\frac{a - x}{a} - g(a, b, t)\frac{x}{a} \right] \frac{y}{b}$$

and verify that $v = g$ on ∂D. Then $w = u - v$ again will be the solution of a Dirichlet problem with zero boundary data, but with a new source term

$$G(x, y, t) = F(x, y, t) - \mathcal{L}v(x, y, t)$$

and the new initial condition

$$w(x, y, 0) = u_0(x, y) - v(x, y, 0).$$

For general boundary data of the third kind one can treat t as a parameter and solve

$$\Delta v = 0$$

$$c_1 \frac{\partial v}{\partial n} + c_2 v = g(x, y, t).$$

Then $w = u - v$ will be subject to homogeneous boundary data, a new source term

$$G(x, y, t) = F(x, y, t) + v_t(x, y, t)$$

and the initial condition

$$w(x, y, 0) = u_0(x, y) - v(x, y, 0).$$

Example 9.2 Standing waves on a circular membrane.

It follows from the general theory that the solution of the initial value problem for a clamped circular vibrating membrane

$$\Delta u - \frac{1}{c^2} u_{tt} = 0, \qquad (r, \theta) \in [0, R) \times (0, 2\pi], \quad t > 0$$

$$u(R, \theta, t) = 0$$
$$u(r, \theta, 0) = u_0(r, \theta)$$
$$u_t(r, \theta, 0) = u_1(r, \theta)$$

can be approximated by

$$u_{MN}(r, \theta, t) = \sum_{\substack{n=0 \\ m=1}}^{M,N} \left[\alpha_{mn}(t) u^1_{mn}(r, \theta) + \beta_{mn}(t) u^2_{mn}(r, \theta) \right]$$

where $u^1_{mn}(r, \theta)$ and $u^2_{mn}(r, \theta)$ are eigenfunctions for the Laplacian on a disk found in Example 8.14.

The expansion coefficient $\alpha_{mn}(t)$ satisfies

$$c^2 \delta^2_{mn} \alpha_{mn}(t) + \alpha''_{mn}(t) = 0$$

$$\alpha_{mn}(0) = \hat{u}_{0mn}$$

$$\alpha'_{mn}(0) = \hat{u}_{1mn},$$

with

$$\hat{u}_{0mn} = \frac{\langle u_0, u^1_{mn} \rangle}{\langle u^1_{mn}, u^1_{mn} \rangle}$$

$$\hat{u}_{1mn} = \frac{\langle u_1, u^1_{mn} \rangle}{\langle u^1_{mn}, u^1_{mn} \rangle}.$$

Here $-\delta^2_{mn}$ denotes the eigenvalue corresponding to u^1_{mn} and u^2_{mn}. It follows that

$$\alpha_{mn}(t) = \hat{u}_{0mn} \cos c\delta_{mn} t + \frac{\hat{u}_{1mn}}{c\delta_{mn}} \sin c\delta_{mn} t.$$

The same equations hold for $\beta_{mn}(t)$ except that $u^1_{mn}(r, \theta)$ is replaced by $u^2_{mn}(r, \theta)$ in the initial conditions for $\beta_{mn}(t)$. Thus u_{MN} is the superposition of standing waves oscillating with frequency $\frac{c\delta_{mn}}{2\pi}$ and amplitudes $u^1_{mn}(r, \theta)$ and $u^2_{mn}(r, \theta)$. The amplitude is zero wherever the eigenfunction has a zero. For example

$$u^2_{m0}(r, \theta) = J_0(\delta_{m0} r)$$

is zero at $\delta_{m0} r = x_{k0}$ for $k = 1, \ldots, m$ where x_{k0} is the kth zero of the Bessel function $J_0(x)$. The nodes of the standing waves are circles on the membrane with radius $r = \frac{x_{k0}}{\delta_{m0}}$. For $n > 0$ we see from

$$u^2_{mn}(r, \theta) = J_n(\delta_{mn} r) \cos n\theta$$

that the Bessel function J_n contributes m circular nodes with radius $\delta_{mn} r = x_{kn}$, $k = 1, \ldots, m$, while the factor $\cos n\theta$ contributes nodes of zero amplitude along the rays

$$n\theta = \left(\frac{1}{2} + k \right) \pi, \qquad k = 0, \ldots, \left[2n - \frac{1}{2} \right]$$

where $[a]$ denotes the largest integer less than or equal to a. Similarly, $u^1_{mn}(r, \theta) = J_n(\delta_{mn} r) \sin n\theta$ is the amplitude of a standing wave which vanishes wherever the Bessel function or $\sin n\theta$ is zero. For illustration the nodal lines for $u^1_{33}(r, \theta)$ are shown in Fig. 9.2.

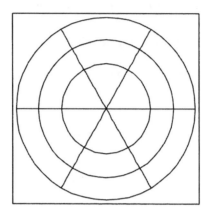

Figure 9.2: Nodes of the standing wave $u^1_{33}(r, \theta)$.

Example 9.3 The potential inside a charged sphere.

Consider the problem

$$\Delta u(r, \theta, \phi) = 0$$

$$u(R, \theta, \phi) = (\vec{n} \cdot \vec{k}_1)^+ + (\vec{n} \cdot \vec{k}_2)^+$$

where $\vec{n}$ is the outward unit normal to the sphere of radius R, $\vec{k}_1$, and $\vec{k}_2$ are two given vectors in $\mathbb{R}_3$, and $a^+ = \max(a, 0)$. In spherical coordinates the formulation is (see Section 1.3)

$$r^2 u_{rr} + 2r u_r + \Delta_{(\theta,\phi)} u = 0$$

$$u(R, \theta, \phi) = (\cos \theta \sin \phi, \sin \theta \sin \phi, \cos \phi) \cdot k^+_1$$
$$+ (\cos \theta \sin \phi, \sin \theta \sin \phi, \cos \phi) \cdot k^+_2.$$

Since we already know the eigenfunctions and eigenvalues for the Laplacian on the surface of the unit sphere, we can write an approximate solution in the form

$$u_{MN}(r, \theta, \phi) = \sum_{\substack{n=1 \\ m=0}}^{\substack{n=m \\ M}} \alpha_{mn}(r) P^n_m(\cos \phi) \sin n\theta + \sum_{\substack{n=0 \\ m=0}}^{\substack{n=m \\ M}} \beta_{mn}(r) P^n_m(\cos \phi) \cos n\theta$$

where

$$r^2 \alpha_{mn}''(r) + 2r\alpha_{mn}'(r) - m(m+1)\alpha_{mn} = 0$$

$$\lim_{r \to 0} |\alpha_{mn}(r, \theta, \phi)| < \infty$$

$$\alpha_{mn}(R) = \frac{\langle u(R, \theta, \phi), P_m^n(\cos\phi)\sin n\theta \rangle}{\langle P_m^n(\cos\phi)\sin n\theta, P_m^n(\cos\phi)\sin n\theta \rangle}$$

with

$$\langle f, g \rangle = \int_0^{2\pi} \int_0^\pi f(\theta, \phi) g(\theta, \phi) \sin\phi \, d\phi \, d\theta.$$

The same differential equation and similar boundary conditions apply to $\beta_{mn}(r)$.

It is straightforward to verify that

$$\alpha_{mn}(r) = \alpha_{mn}(R)\left(\frac{r}{R}\right)^m.$$

Similarly

$$\beta_{mn}(r) = \beta_{mn}(R)\left(\frac{r}{R}\right)^m.$$

The problem is solved once the Fourier coefficients $\alpha_{mn}(R)$ and $\beta_{mn}(R)$ are found. For their calculation the formula

$$\int_0^\pi P_m^n(\cos\phi)^2 \sin\phi \, d\phi = \frac{2(n+m)!}{(2m+1)(m-n)!}$$

may prove helpful [1].

Example 9.4 Pressure in a porous slider bearing.

The following interface problem is a simplification of the mathematical model we found in [17] for the analysis of a tapered porous slider bearing. It is included here to show how the two eigenfunction expansion methods of Example 8.13 can be combined to yield a solution $\{u(x,y), v(x,y,z)\}$ of the following equations:

$$\Delta u + \frac{3}{x}u_x = \frac{B}{x^3} + \frac{C}{x^3}\frac{\partial v}{\partial z}(x,y,1), \quad (x,y) \in D = (x_0, a) \times (0, b), \quad x_0 > 0 \quad (9.1)$$

$$u = 0 \quad \text{on } \partial D$$

$$\Delta v(x,y,z) = 0, \quad (x,y,z) \in D \times (0,1)$$

$$v = 0, \quad (x,y) \in \partial D \text{ or when } z = 0.$$

In addition we require

$$u(x,y) = v(x,y,1), \quad (x,y) \in D.$$

Here u and v are the pressures in the fluid film and in the porous slider. The last term in the Reynolds equation (9.1) models the lubricant flow in and out of the slider as predicted by Darcy's law which governs the fluid flow in the porous slider.

We are going to find an approximate solution $\{u_K(x,y), v_K(x,y,z)\}$ of the form

$$u_K(x,y) = \sum_{k=1}^{K} \beta_k \Phi_k(x,y)$$

$$v_K(x,y) = \sum_{k=1}^{K} \alpha_k(z) \Psi_k(x,y)$$

such that

$$\Delta v_K(x,y,z) = 0$$
$$v_K(x,y,z) = 0 \qquad \text{for } (x,y) \in \partial D \text{ or when } z = 0$$

and

$$\Delta u_K + \frac{3}{x} u_{Kx} = \hat{P}_K \left[\frac{B}{x^3} + \frac{C}{x^3} \frac{\partial v_K}{\partial z} (x,y,1) \right], \qquad (x,y) \in D,$$

$$u_K = 0 \quad \text{on } \partial D.$$

The continuity condition is approximated by

$$v_K(x,y,1) = P_K u_K(x,y).$$

The functions $\{\Phi_k\}$, $\{\Psi_k\}$ and the projections P_K and $\hat{P}_K$ will be introduced below.

Assume for the moment that $u_K(x,y) = \sum_{k=1}^{K} \beta_k \Phi_k(x,y)$ is known. Then v can be found with an eigenfunction expansion in the usual way. We write

$$v_{MN}(x,y,z) = \sum_{m=n=1}^{M,N} \alpha_{mn}(z) u_{mn}(x,y)$$

where $\{u_{mn}(x,y)\}$ are the eigenfunctions of the two-dimensional Laplacian found in Example 8.12, and where P_{MN} denotes the orthogonal projection onto the span of these first MN eigenfunctions. If the corresponding eigenvalues are denoted by $-\delta_{mn}^2$, then

$$-\delta_{mn}^2 \alpha_{mn}(z) + \alpha_{mn}''(z) = 0$$

$$\alpha_{mn}(0) = 0$$

$$\alpha_{mn}(1) = \frac{\langle u_K(x,y), u_{mn} \rangle}{\langle u_{mn}, u_{mn} \rangle}$$

where

$$\langle f, g \rangle = \int_D f(x, y) g(x, y) dx\, dy.$$

The solution is

$$\alpha_{mn}(z) = \frac{\alpha_{mn}(1) \sinh \delta_{mn} z}{\sinh \delta_{mn}}.$$

For the remainder of this example it will be convenient to order the eigenfunctions linearly for $1 \le n \le N$ and $1 \le m \le M$ by defining

$$k = (n-1)M + m, \qquad k = 1, \ldots, K = MN,$$

so that

$$n = [k/M] + 1, \qquad m = k - (n-1)M,$$

and writing

$$\Psi_k(x, y) = u_{mn}(x, y)$$
$$\delta_k = \delta_{mn} \qquad k = 1, \ldots, K = MN.$$

It follows that

$$\alpha_k(z) = \frac{\langle u_K(x, y), \Psi_k \rangle \sinh \delta_k z}{\langle \Psi_k, \Psi_k \rangle \sinh \delta_k}.$$

As a consequence

$$\frac{\partial v_K}{\partial z_K}(x, y, 1) = \sum_{m=1}^{K} D_m \langle u_K(x, y), \Psi_m \rangle \Psi_m(x, y)$$

where

$$D_m = \frac{\delta_m}{\langle \Psi_m, \Psi_m \rangle \tanh \delta_m}.$$

Let us now turn to the solution $u_K(x, y)$. In view of Example 8.5 it is straightforward to verify that the two-dimensional eigenvalue problem

$$\Delta u + \frac{3}{x} u = \mu u, \qquad (x, y) \in D$$

$$u = 0 \qquad (x, y) \in \partial D,$$

has countably many eigenfunctions

$$\Phi_{mn}(x, y) = \frac{1}{x}[Y_1(\lambda_m x_0) J_1(\lambda_m x) - J_1(\lambda_m x_0) Y_1(\lambda_m x)] \sin \rho_n y$$

with corresponding eigenvalues

$$\mu_{mn} = -[\lambda_m^2 + \rho_n^2]$$

where $\rho_n = \frac{n\pi}{b}$ and λ_m is the mth positive root of

$$f(\lambda) = J_1(\lambda x_0)Y_1(\lambda a) - J_1(\lambda a)Y_1(\lambda x_0) = 0.$$

These eigenfunctions are orthogonal over D with respect to the weight function $w(x, y) = x^3$. As above we shall order them linearly and denote by $\hat{P}_K$ the orthogonal projection onto span $\{\Phi_k\}_{k=1}^K$ with respect to the inner product

$$\langle\langle f, g \rangle\rangle = \int_D f(x, y)g(x, y)x^3 dx\, dy.$$

Analogous to our derivation of the Green's function G_{KK} in Example 8.13 we find that

$$u_K(x, y) = \sum_{k=1}^K \beta_k \Phi_k(x, y)$$

is a solution of (9.1) if and only if for $k = 1, \ldots, K$

$$\beta_k = \frac{\left\langle\left\langle \frac{1}{x^3}\left(B + C \frac{\partial v_K}{\partial z}(x, y, 1)\right), \Phi_k \right\rangle\right\rangle}{\mu_k}$$

$$= \frac{\left\langle\left\langle \frac{1}{x^3}\left(B + C \sum_{m=1}^K \sum_{n=1}^K \langle \beta_n \Phi_n, D_m \Psi_m \rangle \Psi_m(x, y)\right), \Phi_k \right\rangle\right\rangle}{\mu_k}$$

$$= \frac{\left\langle\left\langle \frac{B}{x^3}, \Phi_k \right\rangle\right\rangle + \sum_{n=1}^K \sum_{m=1}^K \left\langle\left\langle \frac{C}{x^3} \Phi_k, \Psi_m \right\rangle\right\rangle \langle D_m \Psi_m, \Phi_n \rangle \beta_n}{\mu_k}.$$

These equations can be written in matrix form for $\vec{\beta} = (\beta_1, \ldots, \beta_K)$ as

$$\vec{\beta} = \vec{b} + \mathcal{MB}\vec{\beta}$$

where

$$\mathcal{M}_{km} = \frac{1}{\mu_k}\left\langle\left\langle \frac{C}{x^3} \Phi_k, \Psi_m \right\rangle\right\rangle = \frac{C}{\mu_k}\langle \Phi_k, \Psi_m \rangle$$

$$\mathcal{B}_{mn} = \langle D_m \psi_m, \Phi_n \rangle$$

and

$$b_k = \frac{1}{\mu_k}\left\langle\left\langle \frac{B}{x^3}, \Phi_k \right\rangle\right\rangle = \frac{B}{\mu_k}\langle 1, \Phi_k \rangle.$$

We point out that $u_K(x, y)$ and $v_K(x, y, 1)$ belong to different function spaces because they are expanded in terms of different eigenfunctions. They are linked through the interface continuity approximation

$$v_K(x, y, 1) = P_K u_K(x, y)$$

which is an algebraic system for the $2K$ degrees of freedom in the expansions but does not imply that $v_K(x, y, 1)$ and $u_K(x, y)$ are the same. In view of our numerical experiments with Example 8.11 a least squares minimization of

$$\|v_K(x, y, 1) - u_K(x, y)\|_{L_2(D)}$$

may be a viable alternative for linking u and v at $z = 1$.

9.2   The eigenvalue problem for the Laplacian in $\mathbb{R}_3$

Example 9.5 An eigenvalue problem for quadrilaterals.

Let us consider the representative eigenvalue problem

$$\Delta u = \mu u, \qquad (x, y, z) \in D = D_1 \times (0, c)$$

$$u = 0, \qquad (x, y, z) \in D_1 \times (0, c)$$

$$\frac{\partial u}{\partial n} = 0 \qquad \text{when } z = 0 \text{ or } z = c,$$

where

$$D_1 = (0, a) \times (0, b).$$

We know from Example 8.12 that the eigenvalue problem

$$\Delta \Phi(x, y) = \hat{\mu} \Phi(x, y), \qquad (x, y) \in D_1$$

$$\Phi = 0 \quad \text{on } \partial D_1$$

has the eigenfunctions $u_{mn}(x, y) = \sin \lambda_m x \sin \rho_n y$ with

$$\lambda_m = \frac{m\pi}{a}, \qquad \rho_n = \frac{n\pi}{b}, \qquad m, n \geq 1$$

and eigenvalues

$$\hat{\mu}_{mn} = -[\lambda_m^2 + \rho_n^2].$$

As in Example 8.12 we look for an eigenfunction of the form

$$u(x, y, z) = \alpha_{mn}(z) u_{mn}(x, y).$$

Substitution into the three-dimensional eigenvalue problem shows that

$$\hat{\mu}_{mn} \alpha_{mn}(z) + \alpha''_{mn}(z) = \mu \alpha_{mn}(z)$$

$$\alpha'_{mn}(0) = \alpha'_{mn}(c) = 0.$$

This eigenvalue problem for $\alpha_{mn}(z)$ has solutions

$$\alpha_{mnp}(z) = \cos \eta_p z$$

whenever

$$-\hat{\mu}_{mn} + \mu = -\left(\frac{p\pi}{c}\right)^2 \equiv -\eta_p^2$$

for an integer $p \geq 1$. Hence for any integer $m, n, p \geq 1$ we have the eigenfunction

$$u_{mnp}(x, y, z) = \sin \lambda_m x \sin \rho_n y \cos \eta_p z$$

with corresponding eigenvalue

$$\mu_{mnp} = -[\lambda_m^2 + \rho_n^2 + \eta_p^2].$$

Example 9.6 An eigenvalue problem for the Laplacian in a cylinder.

We consider eigenfunctions which are periodic in z with period c and whose radial derivatives vanish on the mantle of a cylinder with radius R. Hence we consider

$$\Delta_{(r,\theta)}u + u_{zz} = \mu u, \qquad (r, \theta, z) \in D = D_1 \times (0, c)$$

$$u(r, \theta, 0) = u(r, \theta, c), \qquad (r, \theta) \in D_1$$

$$u_z(r, \theta, 0) = u_z(r, \theta, c) \qquad (r, \theta) \in D_1$$

$$\frac{\partial u}{\partial r}(R, \theta, z) = 0, \qquad (r, \theta, z) \in \partial D_1 \times (0, c)$$

where the base of the cylinder is

$$D_1 = \{(r, \theta) : 0 \leq r < R, \quad \theta \in (0, 2\pi]\}.$$

From Example 8.14 we know that the functions

$$u_{mn}^1(r, \theta) = J_n(\lambda r) \sin n\theta, \qquad n = 1, 2,$$

$$u_{mn}^2(r, \theta) = J_n(\lambda r) \cos n\theta, \qquad n = 0, 1,$$

satisfy

$$\Delta_{(r,\theta)}u_{mn}^i = -\lambda^2 u_{mn}^i \qquad \text{for } (r, \theta) \in D_1, \qquad i = 1, 2.$$

The boundary condition at $r = R$ requires

$$\frac{d}{dr} J_n(\lambda R) = 0.$$

For each $n \geq 0$ there are countably many nonzero solutions $\{\lambda_{mn}\}$ such that

$$\frac{d}{dr} J_n(\lambda_{mn} R) = 0.$$

Hence the two-dimensional eigenfunctions are $u^i_{mn}(r,\theta)$ with the corresponding eigenvalues

$$\hat{\mu}_{mn} = -\lambda^2_{mn} = -\left(\frac{x_{mn}}{R}\right)^2$$

where x_{mn} is the mth nonzero root of the function $J'_n(x)$. These roots are available numerically.

When an eigenfunction for the cylinder is written in the form

$$u = \alpha^i_{mn}(z)u^i_{mn}(r,\theta)$$

and substituted into the eigenvalue equation, then $\alpha^i_{mn}(z)$ must be a solution of

$$\hat{\mu}_{mn}\alpha^i_{mn}(z) + \alpha''^i_{mn}(z) = \mu\alpha^i_{mn}(z)$$

$$\alpha^i_{mn}(0) = \alpha^i_{mn}(c)$$

$$\alpha'^i_{mn}(0) = \alpha'^i_{mn}(c).$$

Since $\hat{\mu}_{mn}$ does not depend on i, we can drop the superscript i. Hence we have an eigenvalue problem for $\alpha_{mn}(z)$ which is solved by sines and cosines which are c periodic so that

$$\alpha^1_p(z) = \sin\eta_p z, \qquad p = 1, 2, \ldots$$

$$\alpha^2_p(z) = \cos\eta_p z, \qquad p = 0, 1, \ldots$$

where

$$\eta_p = \frac{2p\pi}{c}.$$

Hence to each of the four eigenfunctions

$$u^k_{mnp}(r,\theta,z) = u^i_{mn}(r,\theta)\alpha^j_p(z), \qquad i,j = 1,2$$

with

$$k = 2(i-1) + j$$

corresponds the eigenvalue μ of the Laplacian

$$\mu_{mnp} = -[\lambda^2_{mn} + \eta^2_p].$$

The eigenfunctions are orthogonal when integrated over the cylinder, i.e., with respect to the inner product

$$\langle f, g\rangle = \int_0^c \int_0^2 \int_0^R f(r,\theta,z)g(r,\theta,z)r\,dr\,d\theta\,dz.$$

Example 9.7 Periodic heat flow in a cylinder.

We consider a model problem for heat flow in a long cylinder of radius $R = 1$ with regularly spaced identical heat sources and convective cooling along its length.

As a mathematical model we shall accept

$$\Delta_{(r,\theta,z)}u - u_t = \delta\left(r - \frac{1}{3}\right)\delta(\theta)\delta\left(z - \frac{1}{4}\right), \quad (r,\theta,z) \in D \times (0,1), \quad t > 0$$

$$u(r,\theta,0,t) = u(r,\theta,1,t)$$

$$u_z(r,\theta,0,t) = u_z(r,\theta,1,t)$$

$$u_r(1,\theta,z,t) = -u(1,\theta,z,t)$$

$$u(r,\theta,z,0) = 0$$

where D is the unit disk.

The approximate solution is written in the form

$$u_Q(r,\theta,z,t) = \sum_{q=1}^{Q} \alpha_q(t)\Phi_q(r,\theta,z)$$

where $\Phi_q(r,\theta,z)$ is a z-periodic eigenfunction of the problem

$$\Delta\Phi = \mu\Phi, \quad (r,\theta,z) \in D \times (0,1)$$

$$\Phi_r(1,\theta,z) = -\Phi(1,\theta,z).$$

Example 9.6 shows that the eigenfunctions are of the form

$$u^k_{mnp}(r,\theta,z) = u^i_{mn}(r,\theta)\alpha^j_p(z)$$

$$k = 2(i-1) + j, \quad i,j = 1,2$$

with u^i_{mn} and α^j_p as given in Example 9.6. The radial eigenvalues $\{\lambda_{mn}\}$ differ from those of Example 9.6. Here λ_{mn} is the mth positive root of the nonlinear equation

$$\lambda J'_n(\lambda) = -J_n(\lambda).$$

For each n there are countably many roots $\{\lambda_{mn}\}$ which must be found numerically. The corresponding radial functions $J_n(\lambda_{mn}r)$ are mutually orthogonal in $\mathcal{L}_2(0,1,r)$. The eigenvalue corresponding to u^k_{mnp} remains

$$\mu^k_{mnp} = -[\lambda^2_{mn} + \eta^2_p].$$

The approximating problem is

$$\Delta u - u_t = \sum_{k,m,n,p} \frac{\langle \delta\left(r - \frac{1}{3}\right)\delta(\theta)\delta\left(z - \frac{1}{4}\right), u^k_{mnp}\rangle}{\langle u^k_{mnp}, u^k_{mnp}\rangle}$$

with the above initial and boundary conditions. It is solved by

$$u_{MNP}^{(r,\theta,z,t)} = \sum_{k=1}^{4}\sum_{m=1}^{M}\sum_{n=0}^{N}\sum_{p=0}^{P} \alpha_{mnp}^{k}(t)u_{mnp}^{k}(r,\theta,z)$$

where

$$\mu_{mnp}^{k}\alpha_{mnp}^{k}(t) - \alpha'{}_{mnp}^{k}(t) = \frac{u_{mnp}^{k}\left(\frac{1}{3},0,\frac{1}{4}\right)}{\langle u_{mnp}^{k}, u_{mnp}^{k}\rangle}$$

$$\alpha_{mnp}^{k}(0) = 0.$$

Hence

$$\alpha_{mnp}^{k}(t) = \frac{u_{mnp}^{k}\left(\frac{1}{3},0,\frac{1}{4}\right)}{\langle u_{mnp}^{k}, u_{mnp}^{k}\rangle}\left(1 - \exp\left(\mu_{mnp}^{k}t\right)\right)$$

for

$$\mu_{mnp}^{k} = -[\lambda_{mn}^{2} + \eta_{p}^{2}], \qquad \eta_{p} = 2p\pi.$$

Note that

$$\langle u_{mnp}^{k}, u_{mnp}^{k}\rangle = \int_{0}^{1} J_{n}^{2}(\lambda_{mn}r)r\,dr\,B_{np}^{k}$$

with

$$B_{np}^{1} = \int_{0}^{2\pi}\sin^{2}n\theta\,d\theta\int_{0}^{1}\sin^{2}\eta_{p}z\,dz, \qquad n,p \geq 1$$

$$B_{np}^{2} = \int_{0}^{2\pi}\sin^{2}n\theta\,d\theta\int_{0}^{1}\cos^{2}\eta_{p}z\,dz, \qquad n \geq 1, \quad p \geq 0$$

$$B_{np}^{3} = \int_{0}^{2\pi}\cos^{2}n\theta\,d\theta\int_{0}^{1}\sin^{2}\eta_{p}z\,dz, \qquad n \geq 0, \quad p \geq 1$$

$$B_{np}^{3} = \int_{0}^{2\pi}\cos^{2}n\theta\,d\theta\int_{0}^{1}\cos^{2}\eta_{p}z\,dz, \qquad n,p \geq 0.$$

Wave propagation in a cylinder is handled analogously. Different boundary conditions at $r = R$ and at $z = 0, c$ will require different eigenfunctions of the Laplacian on a cylinder, but all computations are reasonably straightforward.

Example 9.8 An eigenvalue problem for the Laplacian in a sphere.

In order to solve heat flow and oscillation problems in a sphere we need the eigenfunctions of the Laplacian. We shall study the eigenvalues associated with the Dirichlet problem, i.e.

$$r^{2}u_{rr} + 2ru_{r} + \Delta_{(\theta,\phi)}u = \mu r^{2}u$$

$$u(R,\theta,\phi) = 0, \qquad \lim_{r\to 0}|u(r,\theta,\phi| < \infty$$

where $\Delta_{(\theta,\phi)}$ is the Laplacian on the unit sphere considered in Example 8.15. We know from Example 8.15 that its eigenfunctions are

$$u_{mn}^1(\theta,\phi) = P_m^n(\cos\phi)\sin n\theta, \qquad n = 1,\ldots,m \text{ for } m \geq 1$$

$$u_{mn}^2(\theta,\phi) = P_m^n(\cos\phi)\cos n\theta, \qquad n = 0,\ldots,m \text{ for } m \geq 0$$

with corresponding eigenvalues

$$\hat{\mu}_{mn} = -m(m+1), \qquad m \geq 0.$$

Hence an eigenfunction of the form

$$\alpha_{mn}^i(r)u_{mn}^i(\theta,\phi), \qquad i = 1,2$$

results if $\alpha_{mn}^i(r)$ is a solution of the eigenvalue problem

$$(r^2\alpha')' + (\hat{\mu}_{mn} + r^2\eta^2)\alpha = 0 \qquad\qquad (9.2)$$

$$|\alpha(0)| < \infty, \qquad \alpha(R) = 0$$

where for notational convenience we have set $\mu = -\eta^2$.

This differential equation is known from special function theory (or from (3.10)) to have a bounded solution of the form

$$\alpha_{mn}(r) = \frac{J_{m+1/2}(\eta r)}{\sqrt{r}}, \qquad m = 0,1,\ldots.$$

It follows that the eigenvalues of the Laplacian in spherical coordinates are given by

$$\mu_{\ell mn} = -\eta_{\ell m}^2 = -\left(\frac{x_{\ell m}}{R}\right)^2$$

where $x_{\ell m}$ is the ℓth positive root of the Bessel function $J_{m+1/2}(x)$.

It is customary in special function theory to call the function

$$j_m(z) = \sqrt{\frac{\pi}{2z}}J_{m+1/2}(z), \qquad \text{Arg } z \neq \pi$$

a "spherical Bessel function of the first kind of order m." Similarly, one can define a spherical Bessel function of the second kind of order m in terms of Bessel functions of the second kind

$$y_m(z) = \sqrt{\frac{\pi}{2z}}Y_{m+1/2}(z)$$

and the so-called spherical Hankel function

$$h_m^1(z) = j_m(z) + iy_m(z)$$

$$h_m^2(z) = j_m(z) - i y_m(z).$$

A discussion of these functions and their behavior at $z = 0$ and as $z \to \infty$ may be found, for example, in [1]. In this application then $x_{\ell m}$ is also the ℓth positive root of the spherical Bessel function of the first kind of order m. For each eigenvalue $\mu_{\ell m n}$ with fixed ℓ and m we have the eigenfunctions

$$u_{\ell m n}^i(r, \theta, \phi) = j_m(\eta_{\ell m} r) u_{m n}^i(\theta, \phi), \qquad i = 1, 2, \quad n = 0, 1, \ldots, m.$$

We verify by direct integration over the sphere that the eigenfunctions corresponding to the same eigenvalue are all mutually orthogonal. The general theory assures that eigenfunctions with distinct eigenvalues likewise are orthogonal.

Example 9.9 The eigenvalue problem for Schrödinger's equation with a spherically symmetric potential well.

The last example of this book is a classic textbook problem in quantum mechanics (see, e.g., [20]). It is chosen to remind the reader that eigenfunction expansions are commonplace in quantum mechanics, that the eigenfunctions of the Laplacian are essential building blocks for the eigenfunctions of the Schrödinger equation, and that other eigenvalue problems occur naturally which no longer have the concise structure of Sturm–Liouville problems but which can still be attacked with the algorithms employed throughout these pages.

The arguments of Example 9.8 need only minor modifications when we consider the eigenvalue problem for the Schrödinger equation

$$i\hbar \frac{\partial u}{\partial t} = -\frac{\hbar^2}{2m} \Delta u + V(r)u, \qquad r = \|\vec{x}\| \tag{9.3}$$

where $\hbar$ and m are constants and $V(r)$ is a nonpositive function given as

$$V(r) = \begin{cases} -V_0 & r < a, \qquad V_0 > 0 \\ 0 & r > a. \end{cases}$$

Since V is discontinuous, the above Schrödinger equation cannot have a classical solution everywhere. The postulates of quantum mechanics ask for a function u which is a classical solution where V is continuous, which is continuous and has continuous gradients at all points, and which decays as $r \to \infty$ such that

$$\frac{d}{dt} \int_{\mathbb{R}_3} |u(\vec{x}, t)|^2 d\vec{x} = 0.$$

Let us now solve equation (9.3). An eigenfunction expansion solution would be of the form

$$u(x, t) = \sum \alpha_k(t) \Phi_k(\vec{x})$$

where $\Phi_k(\vec{x})$ is an eigenfunction of

$$-\frac{\hbar^2}{2m}\Delta\Phi + V(r)\Phi = E\Phi$$

with eigenvalue E.

As in Chapter 3 we can multiply this equation by Φ and integrate over $\mathbb{R}_3$ to conclude a priori that any eigenvalue E is real and that Φ can be chosen to be real. The eigenvalue, however, does not need to be positive. In fact, it generally is of interest to discover conditions under which the equation admits negative eigenvalues.

We rewrite the eigenvalue equation in spherical coordinates and obtain

$$\Delta_{(\theta,\phi)}\Phi + (r^2\Phi_r)_r - r^2\tilde{V}(r)\Phi = -r^2\tilde{E}\Phi$$

where $\tilde{V}(r) = \frac{2m}{\hbar^2}V(r)$ and $\tilde{E} = \frac{2m}{\hbar^2}E$. As in Example 9.8 we write

$$\Phi(r,\theta,\phi) = \alpha^i_{mn}(r)u^i_{mn}(\theta,\phi)$$

where u^i_{mn} denotes an eigenfunction of the Laplacian on the surface of the unit sphere with corresponding eigenvalue $\hat{\mu}_{mn} = -m(m+1)$, $n = 0,\ldots,m$. Then $\alpha^i_{mn}(r)$ must be a continuously differentiable solution of

$$(r^2\alpha')' - r^2\tilde{V}(r)\alpha - m(m+1)\alpha = -r^2\tilde{E}\alpha, \qquad 0 < r < \infty.$$

It is subject to the constraints that $|\alpha(0)| < \infty$ and

$$\int_0^\infty |\alpha(r)|^2 r^2\,dr < \infty.$$

This equation is slightly more complicated than (9.2). We consider special cases now.

For $m = 0$ we can write the equation for $\alpha(r)$ as in Example 6.7 in the form

$$(r\alpha)'' + \tilde{E}(r\alpha) = 0, \qquad r > a$$

$$(r\alpha)'' + (\tilde{V}_0 + \tilde{E})(r\alpha) = 0, \qquad r < a.$$

If $\tilde{E} \geq 0$, then we would obtain a linear or trigonometric function for $r > a$ which is not square integrable over (a,∞). Hence any eigenvalue $\tilde{E}$ must necessarily be negative. A square integrable solution for $r > a$ is

$$r\alpha(r) = c\exp\left(-\sqrt{|\tilde{E}|}r\right).$$

The corresponding bounded solution on $[0,a)$ is

$$r\alpha(r) = d\sin\sqrt{\left(\tilde{V}_0 - |\tilde{E}|\right)^1}r.$$

Continuity of $\alpha(r)$ and $\alpha'(r)$ at $r = a$ leads to the equations

$$\begin{pmatrix} \exp\left(-\sqrt{|\tilde{E}|}a\right) & -\sin\gamma a \\ -|\tilde{E}|\exp\left(-\sqrt{|\tilde{E}|}a\right) & -\gamma\cos a\gamma \end{pmatrix}\begin{pmatrix} c \\ d \end{pmatrix} = 0$$

where we have set $\gamma = \sqrt{\tilde{V}_0 - |\tilde{E}|}$. This system can have a nontrivial solution only if the determinant of the coefficient matrix vanishes. After canceling non-zero common factors we find that $\tilde{E}$ must be a negative solution of

$$f(\tilde{E}) = \sqrt{|\tilde{E}|}\sin\gamma a + \gamma\cos\gamma a = 0.$$

For $\tilde{E} < -\tilde{V}_0$, γ is imaginary and there is no root of $f(\tilde{E}) = 0$. $\gamma = 0$ is a root but leads to the trivial solution. For $-\tilde{V}_0 < \tilde{E} < 0$ the sign pattern of the trigonometric polynomial shows that there is an eigenvalue $\tilde{E}_{\ell 0}$ when $(\ell - 1/2)\pi < \gamma a < \ell\pi$. Hence there are at most finitely many eigenvalues $\{\tilde{E}_{\ell 0}\}$ for $m = 0$ with corresponding radial solution

$$\alpha_{\ell m}(r) = \begin{cases} \dfrac{\sin\gamma r\exp\left(-\sqrt{|\tilde{E}_{\ell 0}|}a\right)}{r}, & 0 \le r < a \\[2mm] \dfrac{\sin\gamma a\exp\left(-\sqrt{|\tilde{E}_{\ell 0}|}r\right)}{r}, & a < r < \infty. \end{cases}$$

Because of the exponential decay, it is easy to verify as in Chapter 3 that the set $\{\alpha_{\ell m}(r)\}$ is orthogonal in $\mathcal{L}_2(0, \infty, r^2)$.

Let us now consider the general case when $m \ne 0$. We shall assume first that $-\tilde{V} < \tilde{E} < 0$ and set

$$\gamma = \sqrt{\tilde{V} - |\tilde{E}|}, \qquad \beta = i\sqrt{|\tilde{E}|}.$$

Then the radial equations

$$(r^2\alpha')' + r^2\gamma^2\alpha - m(m+1)\alpha = 0, \qquad r \in [0, a)$$

$$(r\alpha')' - r^2|\tilde{E}|\alpha - m(m+1)\alpha = 0, \qquad r \in (a, \infty)$$

must be solved. Both of these equations are of the form of equation (9.2) with

$$\eta = \gamma, \qquad r \in [0, a)$$

$$\eta = \beta, \qquad r \in (a, \infty).$$

We already know from Example 9.6 that a bounded solution on $[0, a)$ is given by the spherical Bessel function

$$\alpha_m(r) = c_1 j_m(\gamma r).$$

For $r > a$ the solution is a linear combination of the spherical Hankel functions

$$\alpha_m(r) = d_1 h_m^1(\beta r) + d_2 h_m^2(\beta r).$$

The Hankel functions $h_m^1(\beta r)$ and $h_m^2(\beta r)$ decay and increase exponentially for $|\tilde{E}| \neq 0$ as $r \to \infty$. Hence we require $d_2 = 0$ and observe that then $\alpha_m(r)$ belongs to $\mathcal{L}_2(a, \infty, r^2)$. It remains to match the two solutions at $r = a$. As before we need a nonzero solution of the system

$$\alpha_m(a-) = \alpha_m(a+)$$

$$\alpha'_m(a-) = \alpha'_m(a+)$$

or in matrix form

$$\begin{pmatrix} j_m(\gamma a) & h_m^1(\beta a) \\ \gamma j'(\gamma a) & \beta h'^1_m(\beta a) \end{pmatrix} \begin{pmatrix} c_1 \\ d_1 \end{pmatrix} = 0$$

where $'$ denotes the derivative of $j_m(z)$ and h_m^1 with respect to z. $\tilde{E}$ is a possible (scaled) eigenvalue if the determinant of the coefficient matrix vanishes, which leads to the nonlinear equation

$$f_m(\tilde{E}) = j_m(\gamma a)\beta h'^1_m(\beta a) - \gamma j'_m(\gamma a) h_m^1(\beta a) = 0.$$

For $m \geq 1$ this equation can be simplified with the recursion formula

$$g'_m(z) = g_{m-1}(z) - \frac{m+1}{z} g_m(z),$$

which is one of several available for spherical Bessel function.

We obtain after some algebra

$$f_m(E) = \beta j_m(\gamma a) h_{m-1}^1(\beta a) - \gamma j_{m-1}(\gamma a) h_m^1(\beta a). \tag{9.4}$$

It is known that (9.4) has no solution for imaginary γ, that $\gamma = 0$ is a root but leads to the trivial solution, that there are at most finitely many negative roots with their number depending on V_0, and that for positive $\tilde{E}$ the Hankel function solution is not square integrable on (a, ∞). Hence for each $m \geq 0$ there are at most finitely many eigenvalues $\{E_{\ell m}\}$. Their values for a given V_0 are found numerically from (9.4).

If the eigenvalues for a given m are indexed as $\{\tilde{E}_{\ell m}\}$, and the corresponding solutions of the radial equation are labeled $\{\alpha_{\ell m}(r)\}$, then for each such eigenvalue we have the radial eigenfunctions

$$\{\alpha_{\ell m}(r)u_{mn}^i(\theta, \phi)\}, \qquad \ell = 0, 1, \ldots, \quad n = 0, 1, \ldots, m.$$

They are mutually orthogonal in the inner product

$$\langle f, g \rangle = \int_0^{2\pi} \int_0^{\pi} \int_0^{\infty} f(r, \theta, \phi) g(r, \theta, \phi) r \sin \theta dr \, d\theta \, d\phi$$

and can be used to find eigenfunction expansions for the time-dependent Schrö-dinger equation. For example, let $\tilde{E}_{01}$ denote the smallest root of $f_1(\tilde{E}) = 0$. The smallest eigenvalue is then

$$E_{01} = \frac{\hbar^2}{2m} \tilde{E}_{01}.$$

The corresponding solution of the radial equation is

$$\alpha_{01}(r) = \begin{cases} h_1^1(\beta a) j_1(\gamma r) & 0 < r < a \\ j_1(\gamma a) h_1^1(\beta r) & a < r < \infty, \end{cases}$$

and the eigenfunctions for the expansion of solutions of the time-dependent Schrödinger equation are

$$\alpha_{01}(r) u_{11}^1(\theta, \phi) = \alpha_{01}(r) P_1^1(\cos \theta) \sin \theta$$
$$\alpha_{01}(r) u_{10}^2(\theta, \phi) = \alpha_{01}(r) P_1(\cos \theta)$$
$$\alpha_{01}(r) u_{11}^2(\theta, \phi) = \alpha_{01}(r) P_1^1(\cos \theta) \cos \theta.$$

The question of how well an initial condition $u_0(x)$ for the Schrödinger equation (9.3) can be approximated in terms of such eigenfunctions is best left to experts in quantum mechanics, but we do know that if $\{E_k, \Phi_k\}_{k=1}^K$ is a set of K eigenvalues and orthogonal eigenfunctions for (9.3), then the approximating problem

$$i\hbar \frac{\partial u}{\partial t} = -\frac{\hbar^2}{2m} \Delta u + V(r) u$$

$$u(x, 0) = P_K u_0(x) = \sum_{k=1}^K \hat{\alpha}_k \Phi_k(x)$$

with

$$\hat{\alpha}_k = \frac{\langle u_0(x), \Phi_k(x) \rangle}{\langle \Phi_k, \Phi_k \rangle}$$

has the analytic solution

$$u_K(x, t) = \sum_{k=1}^K \hat{\alpha}_k e^{-\frac{iE_k}{\hbar} t} \Phi_k(x).$$

Bibliography

[1] L. C. Andrews, *Special Functions for Engineers and Applied Mathematicians*, Macmillan, ISBN 0-02-948650-5, 1985.

[2] M. Cantor, *Vorlesungen über Geschichte der Mathematik* IV, Teubner, 1908.

[3] R. Courant and D. Hilbert, *Methods of Mathematical Physics* II, Wiley, ISBN 0-471-50439-4, 1962.

[4] H. F. Davis, *Fourier Series and Orthogonal Functions*, Dover, ISBN 0-486-65973-9, 1963.

[5] L. C. Evans, *Partial Differential Equations*, AMS, ISBN 0-8218-0772-2, 1998.

[6] D. Gilbarg and N. S. Trudinger, *Elliptic Partial Differential Equations of Second Order*, Springer, ISBN 3540411607, 2001.

[7] R. B. Guenther and J. W. Lee, *Partial Differential Equations of Mathematical Physics and Integral Equations*, Prentice-Hall, ISBN 0-13-651332-8, 1988.

[8] B. Y. Guo, *Spectral Methods and Their Applications*, World Scientific, ISBN 9810233337, 1998.

[9] E. G. Haug, *The Complete Guide to Option Pricing Formulas*, McGraw-Hill, ISBN 0-7863-1240-8, 1997.

[10] S. I. Hayek, *Advanced Mathematical Methods in Science and Engineering*, Marcel Dekker, ISBN 0-8247-0466-5, 2001.

[11] Abdul J. Jerri, *The Gibbs Phenomenon in Fourier Analysis, Splines and Wavelet Approximations*, Kluwer, ISBN 0792351096, 1998.

[12] E. Kamke, *Differentialgleichungen: Lösungsmethoden und Lösungen I, Gewöhnliche Differentialgleichungen*, Teubner, ISBN 3519020173, 1977.

[13] N. N. Lebedev, *Special Functions and Their Applications*, Dover, ISBN 0-486-60624-4, 1972.

[14] G. M. Lieberman, *Second Order Parabolic Differential Equations*, World Scientific, ISBN 981-02-2883-X, 1996.

[15] D. G. Luenberger, *Optimization by Vector Space Methods*, Wiley, ISBN 0471-55359-X, 1969.

[16] C. R. MacCluer, *Boundary Value Problems and Orthogonal Expansion*, IEEE Press, ISBN 0-7803-1071-3, 1994.

[17] O. Pinkus and B. Sternlicht, *Theory of Hydrodynamic Lubrication*, McGraw-Hill, 1961.

[18] A. D. Polyanin and V. F. Zaitsev, *Handbook of Exact Solutions for Ordinary Differential Equations*, CRC Press, ISBN 1-58488-297-2, 1995.

[19] D. L. Powers, *Boundary Value Problems*, Harcourt, ISBN 0-15-505535-6, 1987.

[20] L. I. Schiff, *Quantum Mechanics*, McGraw-Hill, ISBN 0070552878, 1968.

[21] C. R. Wylie and L. C. Barrett, *Advanced Engineering Mathematics*, McGraw-Hill, ISBN 0-07-072188-2, 1982.

Index

Milton Keynes UK
Ingram Content Group UK Ltd.
UKHW040447071024
449327UK00020B/1046